D0114668

MONSTERS IN THE SKY

MONSTERS IN THE SKY

Paolo Maffei
Translated by Mirella and Riccardo Giacconi

The MIT Press
Cambridge, Massachusetts, and London, England

This book was set in VIP Baskerville by Achorn Graphic Services, Inc.,
printed and bound by The Murray Printing Company in the United States of
America.

Library of Congress Cataloging in Publication Data

Maffei, Paolo.
  Monsters in the sky.
  Translation of the 2d (1977) ed. of I mostri del cielo.
  Bibliography: p.
  Includes index.
  1. Astronomy. I. Title.
QB44.2.M3413   523   79-20041
ISBN 0-262-13153-6

CONTENTS

FOREWORD

It should be made clear at the start of this book that when we speak of monsters of the sky we are not referring to flying saucers, not only because we doubt that they really exist but mostly because we do not consider them sufficiently monstrous. Even people who believe in flying saucers imagine them to be either secret machines from earth or vehicles from outer space. In the first case, they would be no more monstrous than a car or a jet plane; in the second, they would certainly be extraordinary objects, but not beyond our normal understanding. They would be spaceships coming from another planet, built by intelligent beings more technologically advanced than we, yet not too different from the spacecrafts we have built ourselves to land on the moon or to send unmanned probes to Mars and Jupiter. We have no reason not to think that our universe might be inhabited by other intelligent beings, and there are some people who think a meeting with extraterrestrial creatures would be a truly exceptional and wonderful event. The very sighting of UFOs in many cases is nothing more than a projection of this wish.

The monsters we will study are instead true anomalies of the known universe—bodies and phenomena that do not fit into the picture of the universe we have grown accustomed to. After exploring those heavenly bodies familiar to us on a journey in space from beyond the moon to the limits of the universe, we will investigate some extraordinary astronomical objects that were better left out of our initial contact with the extraterrestrial world, either because they could confuse the reader unnecessarily or because they could be considered exceptions. Exceptions, however, are often more important than the rule: what appears to us as a negligible minority may in fact have a determinant effect on the majority, or it may be the first evidence of an enormous group of similar objects that has eluded us in the past, not being as obvious or as easily detectable as most others.

These puzzling objects or phenomena form the subject of this book. Some domestic monsters we will meet are comets that nowadays frighten only superstitious and primitive people, yet still remain even for the scientists abnormal bodies of mysterious origin. We will look at bodies we can see but do not understand, and examine others whose structure and properties we can clearly imagine but cannot prove they exist.

This book is not recommended as a first reading for people who have no idea, or a very vague one, of the world around us and of the heavenly bodies located in it. It is meant principally for those who already have a certain amount of knowledge, albeit elementary, of the structure of the universe and wish to learn more. It is meant for those willing to tackle the investigation of the enigmatic objects that might cause us to reject, or at least thoroughly revise, some of our current ideas and to undertake the study of theories that stem from either the most recent observations or the boldest extrapolations, which construct images close to the realm of science fiction.

Those who have read *Beyond the Moon* will find every subject in this celestial tour dealt with on the basis of what was said in the previous book. The reader who is familiar with the results set forth in those pages is well equipped to understand these. In many cases, precise references have been added that will help the interested reader to take up the subject from the beginning.

The various anomalies we set out to examine are separated into chapters that are not interdependent, and so may be read in any order. They are arranged, however, in such a sequence as to lead the reader from the nearest to the most remote, enigmatic objects. By reading the book from beginning to end, it will be easier to get a panoramic view of an ever-expanding universe where, little by little, even the monsters find their place and justification.

This book requires greater concentration and a more critical approach than that needed to read *Beyond the Moon,* because most of the arguments discussed here have been taken directly from seminal works, and in many cases are still subject to discussion and controversy. No doubt, when all things are explained, a certain overabundance of data and theories will greatly diminish, but for the time being the reader, together with the scientists, has to search for the truth in this plethora of information.

If the space trip *Beyond the Moon* gave the reader the thrill of discovering an unsuspected world, though one already well known to the specialist, perhaps this tour of extraordinary phenomena will help him share in the awe of the scientist confronted with the mysteries of the universe and in the emotion of the researcher who pursues an untrodden path in the hope of finding something unusual just around the next bend. Perhaps a new vista will open up that will give the researcher a complete and harmonious understanding of the field he has investigated for so long; or perhaps he will find a new but nonessential feature, an unsuspected chasm, but, more likely, just another stretch of road where farther ahead a new bend can already be perceived.

I wish to extend my sincerest thanks to those colleagues who have made an essential contribution to the realization of this book by providing a great deal of the illustrative material, generally material that had appeared only in the original scientific publications and in some cases was not readily available or was unpublished. Their names have been gratefully acknowledged in the captions of the respective illustrations.

I am also indebted to Professors Francesco Bertola, Livio Gratton, Franco Occhionero, and Leonida Rosino for reading and discussing parts of the manuscript.

Paolo Maffei

1 THE FASCINATING MONSTERS

The nearest monster can be found beyond the last planetary orbit of the solar system. It moves lazily through the darkness of cosmic space at a distance of about two light-years, half the distance from the earth to the nearest star. It does not shine, since it emits no light of its own, and whatever light it reflects from the surrounding starry sky is so faint that we cannot easily observe it. Should an astronaut approach it, he would detect it only as a dark patch against the brilliant background of the sky—a black shape ever growing until it fills the whole sky. Only by shining a light on it from a very short distance, just as he is about to land, would the astronaut begin to discern its surface, which consists of a glassy, often smooth substance interspersed with fragments of some darker and harder materials. This strange body is less massive than even the smaller planets like earth, and its diameter is only a few tens of kilometers. Too big to be dust, too small to be a planet, too cold and dark to be a star, it would appear to be just a relic from a great cosmic ship-wreck. From its surface one can see the same sky that is visible from earth, but without the moon and the planets. The stars are the same and occupy the same places in the various constellations, but there is one additional star, yellowish and very brilliant (more luminous than Sirius), that shines more vividly than any other on the frozen surface of the body. That star is the sun. Even though it is no more than a speck of light, its attraction is still stronger than that of any other celestial object.

At a certain moment the dark body begins to move more purposefully toward the sun, its speed increasing imperceptibly but continuously: the body has started to fall toward the sun. At first slowly, then faster and faster, it plunges through space in a fall that lasts thousands of years, until one day it reaches the orbit of Pluto, outermost planet in the solar system, crosses it, and penetrates inside the system. By now it has a speed of several kilometers per second, and its fall cannot be stopped; it will travel the entire radius of the solar system in a few tens of years.

As it gets ever closer to the sun, it is no longer compact and dark; its glassy crust begins to reflect the sunlight, and the body finally appears for what it is: an enormous mass of ice mixed with dust, metals, and meteoritic fragments of all kinds. Little by little the ice sublimates, carrying along the lightest particles and forming a sort of atmosphere around the central nucleus, which is still solid. As the distance from the sun diminishes, the temperature of the body rises; the ice melts, vaporizes, and explodes in cavities under the surface; and jets of gas stream out from them to enrich the external atmosphere. Under the influence of solar radiation the body is breaking up, and a cloud of gas (coma) is forming around it. In addition, as the distance gets still smaller, it is also affected by the solar wind which blows the gas away from the coma into a long tail streaming along in the opposite direction from the sun. The monster that thousands of years ago was lurking in the darkness of space is now close to us and appears in our firmament in all its dazzling splendor: it has become a comet (figure 1.1).

We barely have time to see it or photograph it night after night in all its metamorphoses; having completed its lightning journey around the sun, the comet begins to move away, retracting its tail and withdrawing into itself like a cosmic serpent. Ever more slowly but relentlessly, it recedes toward those dark distant regions from which it suddenly sprang to dazzle and awe the mind. Where is it going? Will it come back? When shall we see it again? These are some of the questions left to us earthlings as we wave good-bye to the fascinating monster that is leaving us and will shortly be swallowed up by the depths of space. The monster has gone now. When it cannot be seen any longer, it is quickly forgotten by the man in the street. Astronomers are left with the hard task of answering the questions it raised, not only the ones mentioned but many others concerning its structure and composition that are still under discussion.

Some of the most important questions had already been asked in antiquity, principally by a small number of scholars who correctly believed comets to be celestial bodies. At that time some people regarded them as sinister omens, while others viewed them only as wondrous apparitions. All agreed, however, that they were a phenomenon both unique and mysterious. Comets, to be sure, did not appear to have anything in common with either the fixed stars or the wandering planets. They would appear all of a sudden and then disappear again without a trace. They did not resemble any other celestial objects and even looked different from one another; furthermore, the very same comet would change its shape within the span of a few nights. Finally, they seemed to move in

Figure 1.1   A photograph of comet Ikeya-Seki obtained at dawn on November 2, 1965, by U. Flora of Trieste Observatory with the 76-cm telescope at the high altitude Jungfraujoch Station. At that time the comet's tail was 70,000,000 km long. (Courtesy of M. Hack.)

the most diverse regions of the sky and only rarely remained within the confines of the zodiacal belt where all the planets make their way. For all these reasons, and especially because they looked so different from stars and planets, only a few scholars came to the conclusion that comets were celestial bodies. Most scholars, including Aristotle, believed them to be phenomena of the earth's own atmosphere and sublunar world. Still others viewed them as nothing more than optical illusions without any objective reality. The latter position was reiterated as late as 1623 by Galileo who wrote in the "Saggiatore" that comets were only an illusion created by the effect of sunlight on the thin matter evaporating from earth. Galileo was at the time involved in a sharp controversy with the Jesuit Orazio Grassi, who was a proponent of the theory that comets were of celestial origin. Grassi's view, which would eventually be proved as the correct one, had been first proposed by the ancient Chaldean astronomers and later accepted by the Egyptians, by some Greek scholars, and by the Roman philosopher Seneca. In his work "Naturales Questiones," Seneca explained comets as heavenly bodies traveling in well-defined orbits just like the planets, and he went as far as to prophesy that "one day there shall arise a man who will demonstrate in what regions of the heavens the comets make their way; why they journey so far apart from the other stars, and what is their nature and size." That man would be born sixteen centuries later. His name was Isaac Newton.

By applying the law of universal gravitation that he had discovered, Newton was the first to explain the motion of comets. In collaboration with E. Halley he determined that the comet observed in 1680 had a closed orbit, like that of a planet but much more elongated. Later, Halley applied orbital calculations to twenty-four comets for which sufficiently accurate observations were available and demonstrated that some of their orbits were practically identical. In particular, the orbits of the comets seen in 1682, 1607, and 1531 were in fact one and the same orbit described by a single object passing near the sun and earth approximately every seventy-five years. Through his calculations Halley predicted, as did the French astronomers J. de Lalande and Mme. O. Lepaute, that the comet would reappear in April 1759. Seventeen years after Halley's death, in the spring of 1759, the comet blazed again across the sky. This event proved conclusively that the comet was indeed governed by the same law of universal gravitation as a planet though it might look different from one, and that the comet moved in a closed elliptical orbit of long but finite period. Hence after each revolution it would again become visible to us on earth.

Thus every seventy-five years, Halley's comet starts from aphelion, the point on its orbit which is farthest from the sun and is located beyond Neptune's orbit, and approaches the sun and earth at an ever increasing speed, becoming visible to us only toward the end of its journey. As it moves past the sun, the comet is held captive by its attraction and forced to revolve about it. After passing the point of minimum distance from the sun (perihelion), it begins to recede toward that distant region from which it came, only to start the journey all over again upon reaching aphelion. Since the fateful year of 1759, Halley's comet has already completed two revolutions and is traveling toward its next rendezvous with the sun, which will take place on February 8, 1986.

THE ORBITS

Halley's fundamental contribution was to demonstrate that comets are celestial objects moving in closed orbits, appearing and disappearing periodically not because of a mysterious connection with earthly events but only because their orbits are so elongated. This was a great victory of science over superstition. For science, moreover, it was the first decisive step toward a better understanding of these puzzling objects. Following the example set by Newton and Halley, astronomers proceeded to calculate the orbits of all observable comets to determine in which regions of the sky they moved and with what periodicity they reappeared; whether some of them should be regarded as the same comet in its periodical apparitions; and, finally, which ones could be expected to return in the near future. The problem, at least in principle, is not a difficult one. To determine the position of a comet in space, we only need to know its so-called orbital elements. There are six of them, and what they represent can be easily understood from the definitions that follow and the caption of figure 1.2.

As you will notice, the first two elements, i and Ω, determine the position of the comet's orbital plane in space—its position with regard to the ecliptic plane, or plane of the earth's motion around the sun. The center of the sun, which by definition must occupy one of the two foci of the earth's orbit as well as of the comet's, lies on both planes, that is, it lies on the line of intersection of the two planes, called "line of nodes." The third element, ω, denotes the orientation of the comet's orbit on its plane. The fourth and fifth elements describe the shape of the comet's orbit (circle, ellipse, parabola, or hyperbola). The sixth and last one allows us to calculate at which point of its orbit the comet is to be found at any time, provided we know the time elapsed from perihelion passage and the speed at which it traveled the corresponding segment of its path, which can be derived from Kepler's second law.

It will be noted that the period, P, of the comet's revolution about the sun is not included in the orbital elements. This is due to the fact that it is not an independent variable and can easily be derived from Kepler's third relation once we know the semi-major axis, a.

Once the position of the comet in the sky has been measured on successive nights, we can use the rules of celestial mechanics to derive its orbital elements, in other words, to draw the comet's orbit and find its exact location in space. In point of fact celestial mechanics solves the problem easily when there are only two bodies involved—the sun and the comet—and when we completely disregard the gravitational pull of other bodies, such as that of the planets. This effect, however, can be quite significant when the comet has a close encounter with one of them; in this case the "two-body problem" becomes the "three-body problem" and can be solved analytically only in a few special cases. Moreover, the comet could be affected by more than one planet and, in the extreme case, by all nine of them. In the last instance the problem does not have a general solution; it can be solved numerically for specific cases, and requires very laborious calculations that are made possible today by the use of modern electronic computers.

We shall come back to this subject shortly. For the moment let us examine the results obtained during the two centuries of research after Halley's fundamental discovery. Once the orbital elements of all comets were known, one could determine if there were groups of them that coincided. If so, it followed that some of the comets observed at different times had the same orbit: evidently, one was dealing with a single comet in its periodic returns. Halley's comet is the most famous example of this kind. In addition to this discovery, the study of the orbits led to other important results. The orbital planes of the comets were found to exhibit a wide variety of orientations (figure 1.3). In contrast to the planets, whose motions are nearly confined to the ecliptic plane, the inclination of cometary orbits can be quite large. Pluto, with $i = 17°10'$, has an exceptionally large inclination for a planet, and this is a matter for special study and discussion. Comets, on the other hand, show inclinations up to 90°. Furthermore, many of them do not have a direct or counterclockwise motion, like the planets of the solar system, but rather a retrograde motion. As a result of these findings, astronomers were finally able to explain the appearance of comets near the celestial poles and far from the zodiacal belt, which is centered on the ecliptic, as well as the unpredictable wanderings that had so puzzled the ancients and defeated even Galileo. It still remained a mystery, however, why the orbits of these bodies should have such different orientations in space from all other members of the solar system, despite ever increasing evidence that they, too, belonged to it.

The determination of orbital elements led to another and even more puzzling discovery: not all comets appeared to have elliptical orbits. While many orbits were found to be ellipses of different eccentricity (from very elongated ones to the nearly circular orbit of comet Schwassmann-Wachmann 1), many others turned out to be parabolas or hyperbolas (figure 1.4). This discovery was very important because only comets with elliptical (closed) orbits can be regarded as permanent members of the solar system. All others, according to the geometry of conic curves, should come from infinite distances. In practice, this means that such comets would enter the solar system from stellar space and,

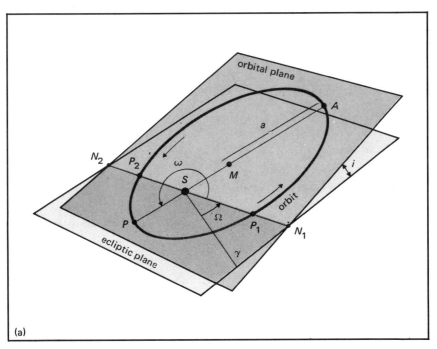

(a)

Figure 1.2 (a) Orbital elements of a typical comet: the reference plane is that
defined by the earth's ecliptic orbit; i is the inclination of the comet's orbital plane
with respect to the ecliptic; Ω is the orientation of the line defined by the intersection of the two planes, called the line of nodes (this angle is measured counterclockwise in the plane of the ecliptic); ω is the angle between the line of nodes
and the line joining the sun to the perihelion (it is measured counterclockwise
on the comet's orbital plane); a is the semi-major axis of the comet's orbit in astronomical units (1 AU = 149,600,000 km); e is the eccentricity of the orbit; and
T is the time of perihelion passage. The first two elements define the position
of the comet's orbital plane in space. The third defines the orientation of the
comet's orbit in its plane. The fourth and fifth characterize the shape and dimensions of the orbit. The sixth gives the date of perihelion passage; from it, knowing the velocity, we can determine the position of the comet at any time.

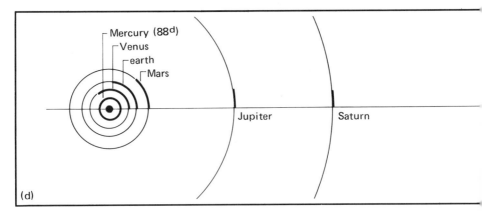

(d)

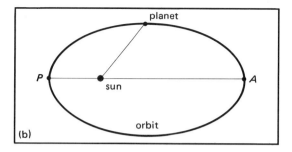

(b) The orbit of a comet, or a planet, around the sun: r is the radius vector, the segment joining the center of the sun to the center of the orbiting body; A is the aphelion, the point of the orbit where the comet (or planet) is farthest from the sun; P is the perihelion, the point where the comet is closest to the sun.

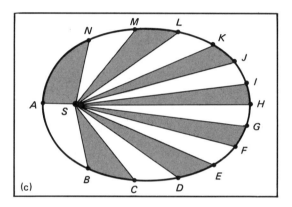

(c) Kepler's laws: (1) The orbits described by the planets around the sun are ellipses in which the sun is at one of the foci. (2) The radius vector sweeps equal areas (the alternating white and gray areas) in equal times. (3) The squares of the periods of revolution of the planets are proportional to the cubes of the semi-major axes. This means that the more distant the planets are from the sun, the slower they move on their orbits, as shown in (d), where the thicker lines represent the portions of the orbits traveled in one mercurian year (88 earth days).

Uranus

Neptune

after shining more or less fleetingly near the sun and earth, would depart never to return.

In discussing this subject, we will make use of the most recent information compiled in B. G. Marsden's *Catalogue of Cometary Orbits*. The catalogue lists the orbits of 964 comets that have appeared from 86 B.C. to March 1975. Of these, 339 correspond to different returns of the same comets—50 of Encke's, 27 of Halley's, 17 of Pons-Winnecke's, for example. Thus the actual number of comets of known orbits observed in more than two thousand years amounts to exactly 625. One hundred and two of these are a group of comets with elliptical orbits and periods shorter than 200 years. The longest period in this group (164.32 years) is that of comet Grigg-Mellish, which appeared in 1742 and again in 1907. From its period and eccentricity we can derive a semi-major axis of 58 AU (astronomical units), which means that this comet reaches just beyond Pluto's aphelion. The other comets in this group have shorter periods and therefore never exceed the confines of the solar system, as defined by Pluto's orbit. As for the remaining 523 comets, the situation is less clear. One hundred and fifty-five of them have orbits that are closely approximated by ellipses and another 85 by hyperbolas, while the remaining 283 are assigned parabolic orbits.

While these conclusions seem fairly clear, simplicity can be deceptive, and astronomers know that any tag put on a comet's orbit oftentimes turns out to be in error. We will become better acquainted with this problem later on. For the moment let us just try to get a general idea of the astronomer's work by following his every step from the first sighting of a new comet.

When a comet is discovered, astronomers try to determine its motion so that they may be able to find it again on the following evenings, and thus measure its exact position at different points in the sky. Using a method first devised by Gauss, only three observations are needed to calculate a preliminary orbit, from which it is then possible to prepare an ephemeris, a table giving the daily positions of a comet and its distances from the sun and earth. After the comet has disappeared, and all available data on its exact positions have been gathered, astronomers calculate a so-called definitive orbit for it, which may turn out to be in the shape of an ellipse, a parabola, or a hyperbola (see figure 1.4). The three cases correspond to values of the eccentricity e smaller than 1, equal to 1, or greater than 1, respectively. When e is very close to unity, it is quite difficult for the astronomer to determine an exact orbital path. An orbit thought to be a very extended ellipse may in fact be an open parabola. Segments of hyperbolic, parabolic, and elliptical orbits coincide almost exactly in the vicinity of the sun (except in the case of short-period comets), as shown in figure 1.4. Unfortunately, when the difference between a parabolic and an elliptical orbit is so small as to fall within the margins of error, mistakes can result in the computation of definitive orbits. At times therefore it is very difficult to decide whether a specific comet moves in an open orbit, which will lead it away from the solar system forever, or whether its path is an enormous ellipse. In the elliptical case the comet will travel to a distance ten times greater than the radius of the solar system. But someday in the distant future, perhaps in hundreds of thousands of years because it is still gravitationally bound to the sun, it will plunge back into the inner part of the system. This could still happen even to comets with hyperbolic orbits, since there is not much evidence to support the notion that the paths are hyperbolic in character. The most clearly hyperbolic orbit among the 85 listed in Marsden's catalogue has been found to have an eccentricity of only 1.004.

The difficulty of resolving open orbital cases is even more obvious if we keep in mind that we must be able to discern an effect that is evident only over great distances, greater than most of the observations at our disposal which refer only to the close neighborhood of the sun. The parabolic orbits attributed to 283 comets might therefore reflect the limitation of our observations rather than actual fact. Another point can be advanced in support of this statement: in the two-body problem a parabolic orbit refers to an object revolving about another of greater mass and coming from an infinite distance where it had zero velocity. A hyperbolic orbit, instead, refers to a body that at infinity had a velocity greater than zero. This is tantamount to saying that in either case the comet would be coming from a place far enough away to be free from the sun's attraction. We see that, whereas the latter case corresponds to a

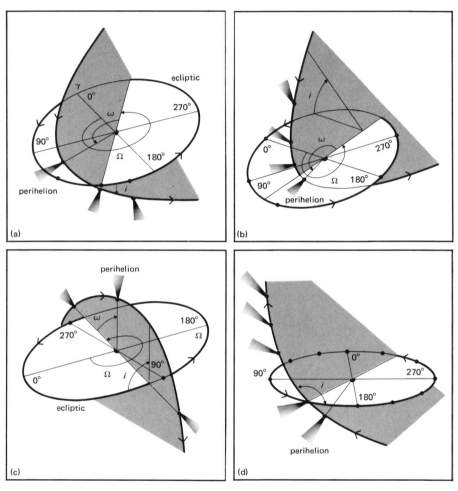

Figure 1.3   Four examples of cometary orbits with different geometric characteristics: (a) comet Friend (1941 a, $i = 26°00'$); (b) comet Cunningham (1940 c, $i = 51°48'$); (c) orbital elements of comet Van Gent-Bernasconi (1941) d, $i = 94°40'$); (d) comet Okabayasi (1940 e, $i = 132°59'$). (Drawings from the originals by A. Fresa published in *Coelum*.)

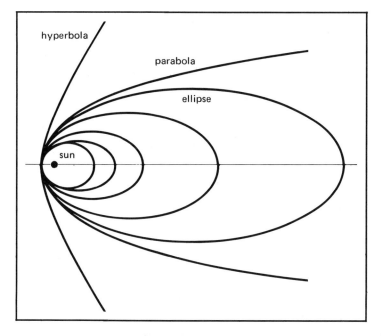

Figure 1.4   Examples of elliptical orbits of different eccentricities: a parabolic orbit and a hyperbolic one. Note that near the sun the orbits nearly coincide.

comet moving from stellar space into the sphere of attraction of the sun with a finite velocity, the former is an extreme case, and presumably quite rare. It is very strange therefore that almost half the orbits listed by Marsden should fall in this category; it would be more reasonable to expect that most, if not all, of the 283 orbits should actually be elliptical or at worse hyperbolic.

Summing up these results, we can conclude for now that only a few comets may not be permanent members of the solar system but a great many certainly are, since their orbits are clearly elliptical. Several other comets are almost certainly bound to the sun, even though they describe orbits of such amplitude that in the computations the orbits are undistinguishable from parabolas. In most cases this uncertainty is due to the fact that the arc of measurable orbit is always too small. The subject, however, is not closed yet; only by taking into account all the forces that act on the comets, in addition to the sun's attraction, will we be able to determine what their orbits are really like and how they might evolve.

SEARCH, DISCOVERY, AND NUMBER OF COMETS

In the past comets were accidentally discovered when they became bright enough to be easily visible to the naked eye, and thus drew the attention of frightened and superstitious people who would rather not see them at all. Today we search for them with a variety of instruments, and we can even seek out the periodic comets by means of long-exposure photographs of the celestial fields where they are expected to appear. Recovery of these comets is made easier when, by knowing their orbits, we can calculate their positions in the sky and detect them as soon as they become observable with the large telescopes, that is, as faint as magnitude 17 or 18. Discovery of all others is much more difficult and takes place in two different ways: either by chance, while we are photographing a certain region of the sky for some other purpose and a comet happens to come into our field of view, or by systematic exploration of the sky with wide-field telescopes of high sensitivity. The chance of discovery is greatest in the regions closest to the sun, shortly after sunset or shortly before dawn, because the comet would then be at the height of its brilliance. Owing to such a constant sky patrol, to which many amateur astronomers also contribute, it is estimated that we can detect all comets that are at maximum brighter than 10th magnitude.

Following discovery, the observer immediately wires the news to a scientific clearinghouse in Cambridge, Massachusetts, which in turn notifies astronomers all over the world by telegrams or special cards that give the position of the comet and, whenever possible, its characteristics and apparent motion on the celestial sphere. The newcomer is temporarily labeled with the name or names (up to three) of the persons who sighted it and the year of the sighting, followed by a small letter indicating the order among all the comets discovered that year. Thus, for example, the famous comet that stirred up so much expectation around Christmas of 1974 bore the name of its discoverer, Kohoutek, followed by the notation 1973 f, indicating that it was the sixth comet discovered in 1973.

From the initial observations astronomers proceed to compute a preliminary orbit that provides the data necessary to follow the comet all through its apparition. They are particularly interested in positional measurements, and try to follow the comet to the limits of their instruments in order to determine its position at the farthest points, which will better define an orbit. Once the definitive orbit has been computed, the comet's designation is changed to the year of its perihelion passage, with a roman numeral replacing the small letter to denote the order of passage. Thus comet Bennett, for example, was temporarily designated as 1969 i and later officially known as 1970 II, since it was the second comet to pass perihelion in 1970.

Knowing the number of comets that were unequivocally observed in the last 2,000 years, we can try to estimate how many comets actually inhabit the heavens. In the past many comets escaped detection for lack of proper instruments. But there is no reason to believe that the number of undetected comets passing near the sun should have changed with time in the last 2,000 years. Thus, considering that in the last 100 years we have discovered an average of four new comets per year, and adopting this number as the mean value also for the past, we find that the number of comets observable with modern instruments from year 1 up to the present must amount to 8,000. This figure, however, is certainly a gross underestimate. In the first place, we cannot observe most comets fainter than magnitude 10 at perihelion nor those whose perihelia lie at solar distances greater than 300,000,000 km, since their orbits are so wide that even when closest to the sun they are still too far away to be observable. Moreover, quite a sizable number of periodic comets must dwell at the present time (as well as in the last 2,000 years) in the farthermost regions of their orbits, which are inaccessible to our observations. Several years ago A. C. Crommelin made an estimate of a total of 160,000 comets in the entire solar system. This estimate may not be too far from the truth. Furthermore, even this number might not remain constant because it seems that "new" comets show up from time to time. The subject of new comets will be considered later, as part of a general accounting. Although it can be proved that comets die, it is not known yet if they are still being born today.

THE REAL ORBITS

ORBITAL PERTURBATIONS
As we have seen, a comet's definitive orbit, when calculated by solving the two-body problem, would correspond to its actual path if there did not exist any other celestial bodies beside the sun and the comet itself. In reality there are 10 major bodies in the solar system (the sun plus 9 planets), and, if we add the comet, we are confronted with the problem of 11 bodies. Unfortunately, although mathematicians have searched for a solution of the n-body problem for over two centuries, a general analytical solution is possible only for $n = 2$, and in very special cases for $n = 3$. To complicate matters, comets have a very small mass, which makes them more susceptible to the attraction of the planets and therefore causes them to deviate from the perfectly elliptical, parabolic, or hyperbolic orbits they would have described had there been only comets and the sun.

Consequently, for the real orbit to be close to the ideal one, the comet would have to pass far from the gravitational pull of all the planets. Naturally, this is an impossible case, since such influence may become very small but never entirely negligible. But should this happen, there would be no appreciable difference between the actual orbit and the one derived by solving mathematically the two-body problem. Often, however, quite the opposite occurs: a comet will pass so close to a planet as to be greatly affected by its attraction. The comet's orbit is then strongly perturbed, and its trajectory may be altered enough to result in a change of orbit. Thus, as a result of a close encounter with a planet, a comet approaching the sun in a parabolic orbit may be deflected into an elliptical one (figure 1.5). In this case the comet is said to have been captured. It is also clear that the reverse can happen: a comet that for centuries has revolved around the sun in an elliptical orbit may be driven by a planet into a parabolic or hyperbolic orbit that will lead it away from the solar system forever.

After the comet has been captured, the aphelion of its new orbit falls near the path of the capturing planet. In the course of time the same thing happens to several other comets, and this results in the formation of a whole family of comets, all having their aphelia near the path of the capturing planet. From this standpoint, the most important planet is Jupiter, heaviest of all, whose family (according to K. Schütte) consists of 52 comets (figure 1.6). Because of their large masses and great distances from the sun, the other outlying planets also contribute to the capture of comets. Saturn's family includes 6 comets, Uranus's 3, Neptune's 8, and Pluto's, perhaps, 5.

To some extent, comet families could reveal the very existence of planets; thus, if the outer planets, from Jupiter on, had never been detected, their presence could be inferred from the discovery of such families.

In a general way, it can be easily understood that the influence of the largest planets is not limited to the capture or expulsion of comets; their gravitational pull is ever present, and particularly intense during the long periods of time when the comets are farthest from the sun and closest to the planets. Proof of this fact has come from a study of 20 cometary orbits carried out at the Institute for Theoretical Astronomy in Leningrad. The Russian researchers have estimated for a period of 400 years (from 1660 to 2060) 140 close approaches to Jupiter, 34 to Saturn, and 2 to Uranus. It appears that in such a time span several comets are drawn into Saturn's family and that Jupiter captures one hyperbolic comet while expelling a former member of its family and deflecting it onto a trans-plutonian orbit. To be precise, a "sphere of influence" can be defined for each planet.[1] Within these spheres the outermost planets act as powerful agents in altering cometary orbits and decisively controlling their evolutions; they can transfer comets from one planetary family to another, and in some cases they can expel them from the solar system or capture them from open orbits.

It should be evident at this point how rudimentary are computations of cometary orbits that are based on observations of very small segments of trajectories. Planetary perturbations show that orbits are never stable; thus at any instant of time a comet's orbit is different from what it was a moment earlier and from what it will be a moment later, if only by a very small amount. Consequently, if we want to determine the true path of a comet as it was before we could observe it, or to predict its future evolution, there is only one way open to us: starting from the observations performed at the time it was visible, we compute a corresponding segment of the orbit and then, working our way backward in time or forward into the future, we calculate what position the comet will occupy in, say, a day's time from the last observation, taking into account both the gravitational attraction of the sun and that of all the planets. From this result we proceed to compute the next day's position, and then we continue building the comet's orbit point by point, taking into account the perturbing effects exerted on the comet at every instant by the prin-

1. The concept of "sphere of influence" was first introduced by P. S. de Laplace. It is defined as that sphere around a planet in which the planet's gravitational attraction is stronger than that of the sun. When a comet enters this sphere, its motion is directed by the planet and the sun becomes the perturbing body. Laplace gave the radius of the sphere of influence the following relationship:

$$\varrho_p = r_p \sqrt[5]{m_p^2},$$

where ϱ_p is the radius of the sphere of influence, r_p the planet's distance from the sun, and m_p the mass of the planet in solar masses. From this formula, we see that the largest sphere of influence is not that of the heaviest planet, Jupiter, but that of Neptune: although less massive than Jupiter, it is farther away from the sun and therefore less influenced by its gravitational pull.

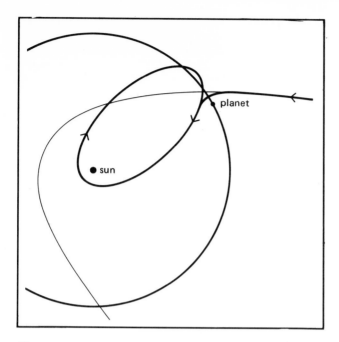

Figure 1.5   Capture of a comet by a major planet. From the original parabolic orbit the comet is transferred onto an elliptical orbit. (From G. de Vaucouleurs, *L'Astronomie.*)

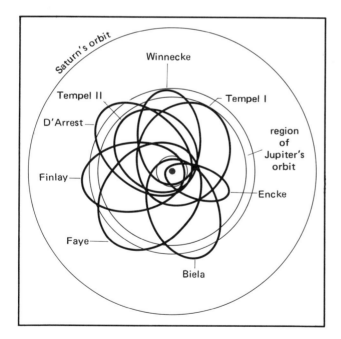

Figure 1.6   The orbits of some of the comets in Jupiter's family, which includes about fifty comets. (From C. Flammarion, *Astronomie Populaire.*)

cipal members of the solar system. The interval that in the above example was assumed to be a day's time, could of course be different depending on the specific case—1 hour, 12 hours, or 10 days, for example.

Such work requires a truly dedicated effort. Nevertheless, not so long ago a few astronomers had the determination to undertake it and the perseverance to carry out a series of repetitive calculations for years on end. Of them we will mention only M. Kamienski, director of the Warsaw Observatory, who devoted forty years of his life to the study of comet Wolf 1 (1884 III); he calculated the evolution of its orbit over a time span of 210 years (from 1750 to 1960), taking into account the perturbations exercised not by one planet alone but by two of them—Jupiter and Saturn—from 1750 to 1884, and by six—Venus, earth, Mars, Jupiter, Saturn, and Uranus—from 1884 to 1960. His results are clearly shown in figure 1.7.

At the present time, electronic computers are used, so orbital computations have become much faster, and can be applied to all the comets for which sufficient data are available, taking into account not only Jupiter's influence but also that of the massive outer planets, such as Saturn, Uranus, and Neptune. Thus at last astronomers can reconstruct cometary orbits and study their evolution by working back in time, comet by comet; they can recover lost comets and even hope to follow them in time until they can solve the mystery of their origin and demise.

Before discussing this subject at greater length, it might be interesting to stop and consider a few cases. Let us start with the history of one such investigation, a tale well worth relating.

COMET DE VICO-SWIFT

The events we are about to narrate started on the night of August 23, 1844, when the Jesuit De Vico of the Observatory of the Collegio Romano discovered a comet that would become clearly visible to the unaided eye in the first week of September. Night after night, several astronomers measured its successive positions in the sky and computed its orbit. Calculations showed that the comet described an elliptical orbit with a period of 5.28 years. Considering the shortness of its period and the uncommon brightness of the object, a number of astronomers won-

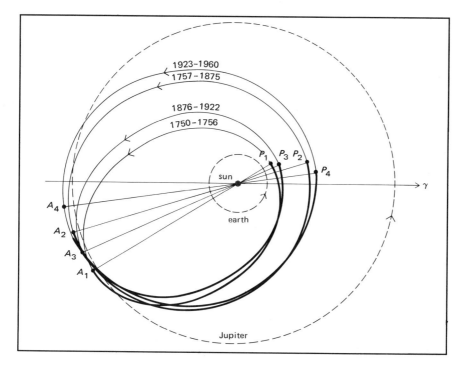

Figure 1.7   Evolution of the orbit of comet Wolf I from 1750 to 1960, accord-ing to calculations by M. Kamienski who devoted forty years to this research. (From *L'Astronomie.*)

dered whether the comet might not have been seen previously. This controversial question was resolved by Le Verrier (the famous astronomer who discovered the planet Neptune by mathematical deduction alone). He suggested that the comet could be the same as that discovered by la Hire on September 11, 1678, and easily observable with the naked eye until October 7 of the same year.

Astronomers attempted to see it again on its subsequent approaches to the sun and earth, but without any success, and, when the comet once again failed to return in 1866, it was considered lost forever.

On the night of November 20, 1894, Edward Swift (son of Lewis Swift, the famous comet hunter) accidentally discovered a comet that, though very faint, was judged to be the same as De Vico's. Its elliptical orbit turned out to be larger than that of De Vico's comet, and its period longer (5.80 years). Nevertheless, the coincidence of the two objects was confirmed, and furthermore the small difference between the two orbits proved to be the explanation of its temporary disappearance. L. Schulhof, who had computed the new period of 5.8 years, found that in 1855 De Vico's comet had undergone a close approach to the planet Jupiter, which had resulted in the lengthening of its period and the widening of its orbit. The latter event caused the comet to move in a path a little farther away from the earth than before, and consequently to become so much fainter that it could not be observed again for many years.

After its rediscovery in 1894, the comet was observed until the end of January 1895. From the new set of measurements the astronomer F. H. Seares computed a period of 5.855 years and predicted that in 1897 the comet would approach to within 64,000,000 km of Jupiter. This event would serve to widen its orbit again, thus substantially increasing the minimum distance between the object and the earth. This further widening of the comet's orbit would result in a longer period of 6.40 years. On the next perihelion passage in 1901, the position of the comet was not favorable to observation. The 1907 passage should have been more favorable, but the comet could not be found. As a result observations of the comet ground to a halt.

In 1963, B. G. Marsden set out to search for some of the lost periodic comets by means of the mathematical technique previously described.

Starting from the last orbit computed for comet De Vico, he retraced its path, step by step, for all subsequent years, taking into account the perturbations caused by Venus, earth, Mars, Jupiter, and Saturn. Thus he confirmed that comets De Vico and Swift were indeed the same object. Furthermore, by extending his calculations to 1975, he discovered that the widening of the orbit reported by Seares was still taking place. Another close approach to Jupiter, expected to occur in 1968, would increase the comet's perihelion distance to 22,000,000 km and its period to over seven years.

The last favorable opportunity to observe the object, now very faint, came in the summer of 1965. Fortune smiled on this last rendezvous. The search, based on the new computations, was successful, and on the night of June 30, 1965, the missing comet De Vico-Swift was at last located and photographed. Seventy years had passed since it had been last observed, and during that period it had approached the earth no less than ten times.

Photographs taken at the Argentine station of Yale Observatory with the 50-cm double refracting telescope, showed the comet to be of 17th magnitude, almost a hundred times fainter than when it was last seen.

After the 1968 encounter with Jupiter, the comet was deflected so far from earth that even under the most favorable conditions it will never be brighter than 18th magnitude. Thus we will no longer be able to detect it with the telescopes currently available. It is true that we have powerful telescopes to observe objects even a hundred times fainter than the De Vico-Swift comet, but only if they are pointlike, such as stars, and not diffuse and nebulous like comets.

As a result of the constant widening of its orbit caused essentially by Jupiter's perturbing influence, a comet that two centuries ago was bright enough to be easily visible to the naked eye has grown progressively fainter, to the point of vanishing altogether into a darkness that no man can penetrate.

The history of this discovery, brought about by a close cooperation between computational and observational work, is not unique. The same technique has been successfully used to recover several lost comets: for example, comet Holmes, observed in 1892, 1899, and 1906, and rediscovered in 1971; comet Shayn-Schaldach, discovered in 1949 and re-discovered in 1971; and comet Temple 1, observed in 1867, 1873, and 1879, which, after two close approaches to Jupiter, underwent a marked alteration in its orbit and was eventually recovered on January 11, 1972.

COMETS WHIPPLE, BROOKS, AND LEXELL

Several other cases, even more complex and spanning wider time intervals, were investigated and solved by H. I. Kazimirtchak-Polonskaya in Leningrad. His findings have given us a comprehensive view of the vicissitudes three periodic comets—Whipple, Brooks, and Lexell—have undergone in the past three centuries, or will undergo in the next.

Kazimirtchak-Polonskaya traced comet Whipple back to 1660; at that time it was a member of Saturn's family of comets, and so it remained for the next one hundred and fifteen years (figure 1.8). The comet's perihelion distance was then over 5 AU, so it could not be seen from earth even under the most favorable conditions. Following several close approaches to Jupiter, the comet was captured by the planet. Thus its aphelion was drawn near Jupiter's orbit, and its perihelion came to be 2.2 AU from the sun. Between 1915 and 1925 the comet settled into an even tighter orbit, closer to the sun and earth, where it will remain until 1979. Because of these circumstances the comet could now be seen, the first sighting being that recorded by Whipple in 1933. Since then it has been observed five times. However, the comet will undergo two close and long-lasting approaches to Jupiter in the years 1979 to 1985 and 2039 to 2046. Its orbit will again become wider, and its perihelion distance greater (3.8 AU), so once more the comet will be undetectable from earth.

In the period 1660 to 1773, comet Brooks moved in a very wide orbit, with its perihelion lying near Jupiter's path and its aphelion beyond Saturn's (figure 1.9). As a result it was not observable from earth. In 1733 the comet approached to within 0.86 AU of Jupiter; this close passage had the effect of lengthening its path and moving its aphelion near the half-way point between the orbits of Saturn and Uranus. This orbit remained stable for more than a century and a half, but around 1885 the comet started to move closer and closer again to Jupiter, and its aphelion

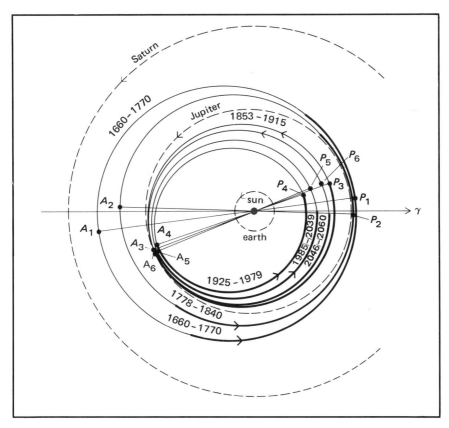

Figure 1.8   Evolution of the orbit of comet Whipple and its transfer from
Saturn's to Jupiter's family. The comet had a close approach to Saturn in 1710
and three passages of long duration from 1770 to 1778, 1840 to 1853, and 1915
to 1925. In 1922 the comet entered the sphere of influence of Jupiter, approaching the planet to within 0.254 AU. In 1979 to 1985 and 2039 to 2046, it will make
two more long-lasting approaches to Jupiter. (From *L'Astronomie*.)

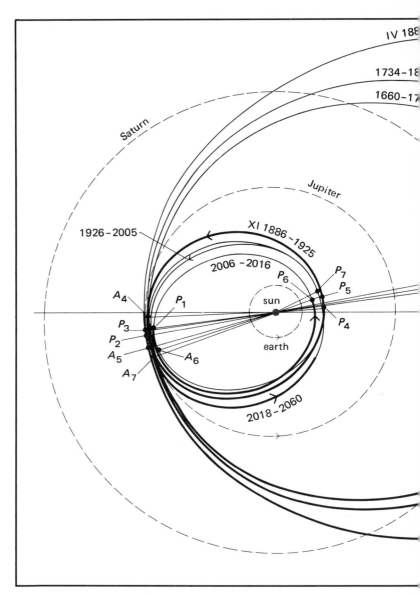

Figure 1.9   Evolution of the orbit of comet Brooks. In 1886 Jupiter captured it from Uranus's family. (From *L'Astronomie*.)

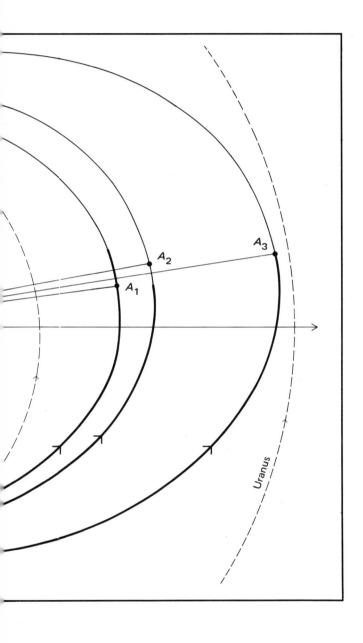

was displaced progressively closer to Uranus's path. By April 1886 the comet clearly belonged to Uranus's family. As it happened, the comet traveled only a small segment of this new and wider orbit because in 1886 it penetrated deeply into Jupiter's sphere of influence, approaching the planet within 0.00096 AU. This close encounter with Jupiter had catastrophic effects: the ascending and descending nodes exchanged places so that the arc of orbit previously above the mean plane of the solar system ended up underneath, and vice-versa; the new orbit shortened so much that the perihelion of the old orbit became the aphelion of the new one, and the new perihelion distance (the minimum distance from the sun) decreased from 5.45 to 1.95 AU. Lastly, the period of revolution, which previously had been longer than forty years, shortened to only seven years. Following this radical alteration of its orbit, which occurred in September 1886, the comet began to pass much nearer to earth, and thus could be discovered by Brooks on its first close approach in 1889. Comet Brooks still is in the family of Jupiter, where it will remain, except for minor orbital perturbations, at least until 2060, the entire period for which computations have been carried out. However, because its aphelion lies near Jupiter's path, sooner or later in the next century a new encounter with this massive planet will transfer the comet into an entirely different orbit.

We come now to the fascinating story of the third comet. On June 14, 1770, C. Messier, the famous comet hunter, discovered a comet that by the end of the month had become bright enough to be easily visible to the unaided eye. Several measurements were taken during the entire time it was visible, and its orbit was computed. At that time Halley's was the only known periodic comet with a clearly elliptical orbit. All other cometary orbits were computed by approximating them to parabolas, a solution that up to that time had appeared quite satisfactory. The new comet's orbit, however, did not seem to fit any parabolic description. Computations by A. Lexell showed that the path was closer to an ellipse with a period of five to six years. Apart from this novelty, what greatly puzzled astronomers was how a comet with such a short period and so easily visible to the naked eye could have gone undetected during all its previous perihelion passages. After a further detailed analysis,

Lexell was able to answer this objection. He noted that in 1767 the comet had sustained an encounter with Jupiter that had substantially reduced its orbit. Prior to that year, its perihelion distance was 3 AU, and the comet could not be seen from earth even on its closest approach to the sun. In 1767, however, the orbit had shortened, the perihelion distance had decreased to only 0.7 AU, and the comet could then be discovered from earth. On its 1776 return observing conditions were not favorable, and three years later the comet had a second encounter with Jupiter which deflected it into a new orbit so distant from earth that it could no longer be seen. Lexell's revolutionary ideas did not conform with the views generally held at the time and were the subject of debate and criticism for over a century. Recent studies made by Kazimirtchak-Polonskaya, spanning four centuries, have completely vindicated Lexell's views (figure 1.10). Quite rightly, the comet bears his name today, even though it was actually discovered by Messier.

These tales, and others that will shortly be told, demonstrate how difficult it is to try to categorize comets according to standard patterns of behavior. Different ways of interacting with other celestial bodies coupled with structural differences make each comet a unique case. Every comet, like every man, has its own history.

THE NATURE OF COMETS

NUCLEUS, COMA, AND TAIL
It is actually the nucleus, the densest and heaviest part of the comet, that moves along the types of orbit we have described so far. Coma and tail develop around the nucleus (figure 1.11), but the matter dispersed in the coma and tail pursues different orbits, whose shapes are influenced also by forces other than gravity.

The nucleus and coma together form that part of the comet commonly known as the head. Though the densest and most important part, the nucleus is so small as to be practically invisible even by telescope. On May 19, 1910, Halley's comet passed directly in front of the sun's disk. If the nucleus was at least 16,000 km in diameter as it was estimated, it should have been visible as a small black dot against the surface of the sun. Despite the most accurate observations, nothing could be seen, and this proves that the nucleus must have been much smaller. More recently, photometric measurements performed on a dozen comets have given values for the radii of the nuclei ranging from about 100 m to about 50 km.

Despite its smallness, the nucleus is the most important part of the comet because most of the comet's mass is concentrated in it. Particles and gases, which are continuously released from the nucleus, diffuse into space to form the coma and tail, both of which are much brighter than the nucleus though much less dense.

In contrast to the small size of the nucleus, the dimensions of the tails are enormous: they often are of the order of some 10,000,000 km. Halley's comet of 1910 exhibited a tail 110,000,000 km long; the comet of 1811 reached 176,000,000; and that of 1860 is estimated to have been at least 240,000,000 km. The record is held by the comet of 1843 which attained a length of 320,000,000 km: it was so long that when its head was near the sun, the last tenuous tendrils of its tail still brushed Jupiter's orbit.

Cometary tails differ substantially not only in length but also in structure, shape, luminosity, and number. It is not uncommon to find comets with multiple tails. A celebrated example is that of comet Chéseaux of 1744, which exhibited no less than six tails in a fan shape 44° wide and 30° to 40° long.

Notwithstanding their enormous sizes, comets are bodies of very low density and mass. The fact that they are extremely tenuous can be appreciated even by direct observation. As already noted by the ancients —notably Democritus, Aristotle, and Seneca—comets are completely transparent when viewed against the stars. Even when a star is concealed by the head, which is the densest part of the comet, there is no perceptible dimming of its brightness.

The values found for cometary masses are also exceedingly low. Not even during the closest encounter with a planet does a comet cause the least variation in its motion. On July 1, 1770, Lexell's comet passed only 2,500,000 km from the earth. Had it been as massive as the earth, our planet would have been deflected from its path onto a slightly wider

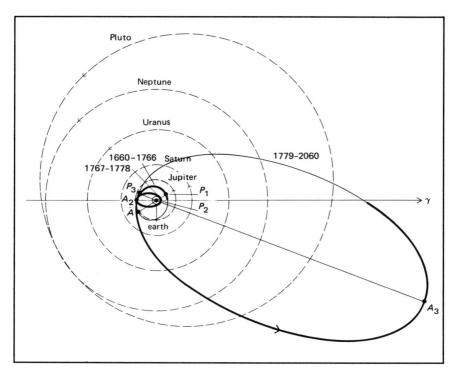

Figure 1.10   Evolution of the orbit of comet Lexell between 1660 and 2060. In 1779 it was expelled from Jupiter's family and transferred onto a trans-plutonian orbit. (From *L'Astronomie*.)

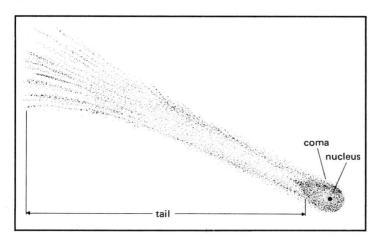

Figure 1.11   The structure of a comet: the coma envelops a compact, small nucleus (tens or hundreds of kilometers); solar wind pressure shapes a long tail of gas and dust.

orbit, and our year would have become longer by 2 hours and 47 minutes. None of this happened, and from this fact Laplace deduced that the mass of the comet was necessarily less than 1/5000 that of the earth. It is impossible therefore to compute the mass of a comet from the perturbations it might cause in the motion of the planets or satellites, all of which are much too heavy to be affected in any way. However, there have been instances of comets splitting up into two or more parts, and in such cases it is possible to study their mutual perturbations and thereby derive the mass. Comet Biela, for example, was found to have a mass one ten-millionth that of the earth and comet Wirtanen one ten-billionth. These values are extremely low by planetary standards but actually correspond to 600,000 billion tons and 100 billion tons, respectively.

CHEMICAL COMPOSITION

After measuring and weighing these strange celestial objects, we shall now examine their chemical composition and physical state. Spectroscopic analysis of the comets began in the second half of the nineteenth century and in recent years has become a very powerful tool for their study. In particular, it brought about the discovery of gaseous molecular compounds in the cometary head, such as C_2, CN, CH, NH, OH, as well as the identification of the ionized molecules CO^+ and N_2^+ in the tail. It is still under discussion whether triatomic molecules like NH_2 and C_3 are present in the head. The spectra of comets closely approaching the sun often exhibit the D lines of sodium, as well as the lines of other metals, notably, iron and nickel. In addition to a band emission spectrum due to excitation phenomena by solar radiation, there is also a continuous spectrum of reflected sunlight (figure 1.12). The former is produced by gases, the latter by dust. Since the second type of spectrum is at times very intense and at other times almost nonexistent, there can be comets rich in dust as well as others containing only gases.

Another characteristic is worthy of note: since the tail always forms on the side opposite the sun, a comet moves head first when approaching the sun and tail first when receding. In comets with two tails (figure 1.13) observations have shown that one of them is exactly opposite the sun while the other lags behind with respect to the direction of motion. The

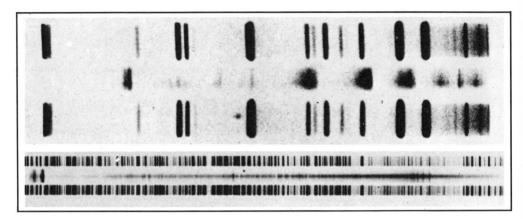

Figure 1.12 The spectra of two comets: *Above,* the band spectrum of comet
Encke obtained on January 14, 1961, by P. Maffei with the 120-cm telescope at
Asiago. The spectrum is at low dispersion. *Below,* the spectrum of comet Arend-
Roland obtained on May 7, 1957, by G. Mannino with the same telescope at
much higher dispersion. *Center,* in both cases the comet spectrum is bracketed
by two laboratory reference spectra. The spectrum of comet Arend-Roland
shows an intense continuum indicative of dust, while that of comet Encke shows
only the bands due to gas.

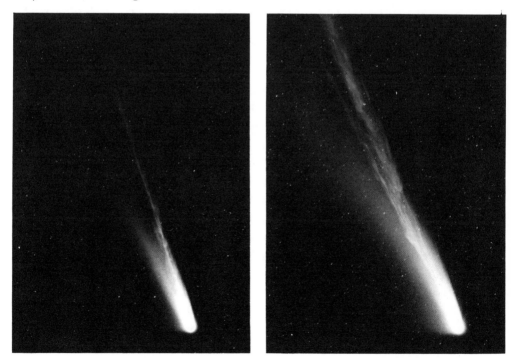

Figure 1.13 Two photographs of comet Mrkos obtained on August 26 and 27,
1957, respectively, with the 122-cm Schmidt telescope at Mount Palomar. Note
the evolution of the tail within 24 hours. (The Hale Observatories.)

former has been found to consist only of gases and the latter of dust. This is due to the fact that tails are formed by the direct action of solar wind and radiation pressure on the particles expelled from the nucleus, a pressure that is stronger on gases than on the heavier dust particles. Further proof that comets are "fashioned" by the sun is provided by the long-known fact that at great distances from the sun, that is, at more than 1.5 AU, comets have no tail. Thus their appearance is not at all nebular, and they can be mistaken for stars.

The molecules found in comets cannot be easily seen under laboratory conditions, since they are unstable chemical radicals produced only by the dissociation of more complex stable molecules. Considering that they are the basic constituents of the comets, it may be interesting to examine these original molecules, or "parent molecules." Almost certainly, they are the molecules CH_4 and NH_3, both of which must be present in the nucleus in the solid state, in the form of ice. Recently, it has become increasingly evident that the nucleus must contain significant amounts of H_2O ice. The presence of frozen water, which was required in any case to explain the origin of the radical OH^+ observed in the tail, has been confirmed by two important discoveries.

Until three years ago, the study of comets was strictly earth-bound: all observations—visual, telescopic, photographic, and spectrographic— were carried out only from the surface of the earth. In January 1970 a comet (Tago-Sato-Kosaka) was observed for the first time from space by means of the artificial satellite OAO-2. An investigation of the comet's ultraviolet spectrum, made possible by the fact that the satellite was operating outside the earth's atmosphere, led to the discovery of a feature due to the radical OH, and of a strong hydrogen emission at the L_α line. This was the first time that free hydrogen had been detected in a comet. Even more remarkable, however, was the discovery that the hydrogen cloud extended out one and a half degrees from the nucleus: considering the comet's distance at the time of observation, this corresponded to 1,700,000 km. This envelope, then, was truly enormous. At the same time, telescopic observations showed the head to be only 8 arc minutes. This meant that the head of the comet, whose diameter in the visible was 150,000 km, was surrounded by a hydrogen cloud larger than the sun. Soon after, the satellite OGO-5 observed another comet (Bennett) and revealed that it was surrounded by an even larger hydrogen cloud, 10,000,000 km at the point of maximum width and 14,000,000 km in length. The same thing happened in the case of comet Kohoutek (figure 1.14).

Such observations have shown that comets are enormously larger than they appear on earth to the naked eye or through the telescope, and, more important, that hydrogen and the radical OH are much more abundant than previously believed. It follows that, when atoms of hydrogen combine with molecules of the radical OH to form the parent molecule H_2O, and when the comet's distance from the sun is so large that water turns to ice, the resulting amount of solid matter must be enormous and certainly must constitute a high percentage of the nucleus.

In spite of these findings, the H_2O molecule itself was not directly observed until the end of 1973, when molecular bands due to ionized H_2O were independently discovered in comet Kohoutek (figure 1.15) at the Lick and Asiago Observatories. The presence of water in comets, which had long been suspected, was thus clearly ascertained for the first time. A few months later this discovery was confirmed further by the identification of H_2O^+ in the tail of another comet (Bradfield, 1974 b). Observations have shown that the variations of intensity in these bands at different distances from the sun are very similar in both comets.

MODELS OF THE NUCLEUS

These recent discoveries have given strong confirmation to the model of the nucleus that F. L. Whipple had proposed as early as 1950.

According to his interpretation—which we have tacitly adopted from the beginning—a comet, when very far from the sun, consists of a solid block in which all the gases that will later develop near the sun are frozen in the chemical state of the parent molecules. This block is a conglomerate of various ices (CH_4, NH_3, H_2O, and possibly C_2N_2) and nonvolatile meteoritic material, such as silicates, oxides, carbon particles, and metals. As the comet approaches the sun, a great part of the ice sublimates at an ever-increasing rate, carrying along some of the solid particles. In the

Figure 1.14 Two images of comet Kohoutek at the same distance: the one above from earth in visible light, the one below from space in ultraviolet L_α radiation. (The Johns Hopkins University; U.S. Naval Research Laboratory.)

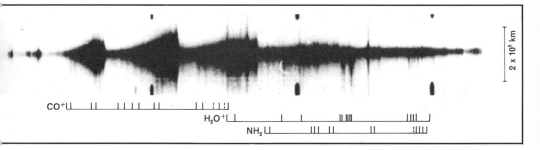

Figure 1.15   A photograph of comet Kohoutek obtained on January 17, 1974, with the 90/65 cm-Schmidt telescope at Asiago and the comet's spectrum. The black bar in the photograph indicates the region from which the spectrum was obtained by P. Benvenuti with the telescope at Cima Ekar. The lower portion of the spectrum corresponds to the region toward the tail; it is in this direction that one finds the ionized molecule of H_2O. (Courtesy of P. Benvenuti.)

comets that pass closest to the sun, molecules and particles escape from
the nucleus at speeds of several tens of meters per second. This rapid
loss of material could well destroy a comet after a few close approaches
to the sun but for two mechanisms that protect the comet. First of all,
vaporization of the ices uses up a great deal of the sun's energy,[2] and
therefore the temperature of the nucleus rises by a relatively small
amount. Second, after the outer layers have been vaporized, the meteoritic material that has not been carried away, and is now free from
ice, forms a spongy and poorly conductive layer that protects the inner
ices.

Whipple's theory, known for obvious reasons as the "dirty snowball"
model, was rivaled until a few years ago by the so-called "gravel-bank"
theory. In this model the cometary nucleus would consist entirely of
solid particles, from which there develop the gases observed in the coma
and tail. The most recent observations, however, have provided strong
evidence against this model, confirming instead the dirty snowball
theory which by now has become generally accepted by all astronomers.

In addition, Whipple's theory explains another important discovery of
recent years. It is a well-known fact that some comets do not follow the
computed orbits exactly, even after allowances are made for planetary
perturbations. A good example is comet Encke, named after the German
astronomer who studied it at the beginning of the nineteenth century.
This comet, which has the shortest known period (3.3 years), would pass
perihelion two and a half hours earlier than predicted. This effect is still
evident today, though to a lesser degree. Halley's comet, according to recent studies of T. Kiang, returns to perihelion with an average delay of
4.1 days. This strange phenomenon seems to be rather widespread. According to B. G. Marsden and Z. Sekanina, only two out of the twenty
periodic comets they have studied from this viewpoint show no deviations; all the others are either early or late. These results show that cometary nuclei are not subject to gravitational forces alone but are also
affected to varying degrees by other forces, whose origin can be found in
the nucleus itself.

2. Sublimation heat is very high: 103 cal/g for C_2N_2 and 670 cal/g for H_2O.

Let us assume for a moment that the nucleus undergoes a violent explosion under the influence of solar energy, a not impossible occurrence. In view of the small mass of the nucleus, such an explosion could impart to it a sudden push in the opposite direction, and the comet would then behave like a spaceship after ignition of a rocket engine. According to Whipple's theory, it is not even necessary to invoke a series of explosions, which would result only in irregular orbital variations. We need only remember that the gases normally ejected at high speed toward the sun impart to the nucleus a continuous push in the opposite direction (figure 1.16a). The deflected nucleus would then move away from the sun into a slightly larger orbit. By itself, this would fail to explain both of the observed cases, but Whipple notes that we need only to assume an axial rotation of the nucleus to understand the situations of orbital deflection shown in figures 1.16b and c. In the case of counterclockwise rotation the jet reaction acts in the forward direction, and the comet accelerates; in the case of clockwise rotation it acts in the backward direction, and the comet undergoes a deceleration.

The only weak spot in this interpretation is that the rotation of the cometary nucleus, which after all is invisible, has never been actually observed. On the other hand, it must be noted that, whenever it could be observed, an axial rotation was indeed found for all bodies in the universe. There is no reason to believe that cometary nuclei are excluded from this seemingly universal rule.

Whipple's model therefore appears quite convincing and suggestive. Apart from this, the study of nongravitational forces has led to other interesting results. In the early 1970s, while studying a number of short period comets, Marsden and Sekanina found that in the course of time (namely, as comets keep on losing mass through repeated approaches to the sun) nongravitational effects would diminish in some cases and increase in others. This observation led them to revise the dirty snowball model. In their view there are two types of cometary nuclei. One type consists of a porous meteoritic matrix that is impregnated with the "snows" of volatile substances (mostly water) and surrounded by an envelope of ice mixed with dust particles. In time these nuclei lose all their volatile material through explosions. Then, since they have only a solid

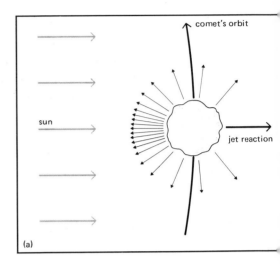

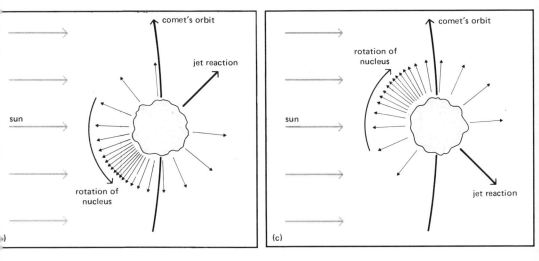

Figure 1.16   A jet reaction model of the nucleus illustrating nongravitational forces produced by the rotation of the nucleus: (a) Solar radiation causes the ices to sublimate and eject molecules (light gray arrows) with a jet reaction effect that pushes the comet's nucleus away from the sun. (b) If the nucleus rotates in the direction of its motion around the sun, the jet reaction pushes the comet away from the sun, resulting in an increase in the comet's period. (c) If the nucleus rotates in the opposite direction, the comet decelerates. (From F. L. Whipple, *Scientific American*, February 1974.)

matrix left, they can no longer be affected by nongravitational forces. The second type of cometary nucleus does not have a meteoritic matrix but consists only of volatile material and free dust particles. As such nuclei lose more and more mass with the passage of time, they are increasingly wrecked by nongravitational forces and eventually disintegrate completely.

MYSTERIES

From all we have said one might conclude that every aspect of comets has been satisfactorily explained, at least in a general way, and that only the detailed investigation of these bodies in all their individual differences can lead to important new discoveries. This is not true. Discoveries and models, such as we have described, give us a satisfactory overall picture but leave several questions still unresolved. Apart from the mystery of their origin, which we shall confront later on, there are other unexplained features in the behavior of comets. Let us consider, for example, the variations in brightness.

A well-publicized case is that of comet Kohoutek (1973 f). At the time of the comet's discovery, its behavior indicated that it would become extremely bright in the vicinity of the sun and earth. Its arrival was awaited in a fever of anticipation that changed to great disappointment when the comet turned out to be barely visible with the naked eye. Although most people have remained with the impression that scientists were completely fooled by comet Kohoutek, what actually happened was not quite that simple, as we shall soon see. Similar baffling events had occurred previously. On September 5, 1940, L. Cunningham discovered a comet that was rather faint but from initial observations appeared to be brightening so rapidly that at the time of perihelion passage it could be expected to be as brilliant as Venus. Despite the fact that war was raging in Europe and people were very grim, the newcomer received a great deal of attention from the press, and many people waited for the comet to burst into brilliance in the sky of Epiphany with a certain apprehension well justified by the times. However, nothing happened out of the ordinary. Starting from the end of October, the increase in luminosity slowed down abruptly, and when the comet finally appeared at year's end, it could be seen only by astronomers and a few amateurs who searched for it with telescopes and field glasses. This example is remarkable for its similarity with the circumstances attending the arrival of comet Kohoutek, yet it is not the most spectacular of all. There have been comets, such as Pajdusakova (1954 II) and Westphal (P), that did not brighten at all upon nearing the sun but kept on fading until they finally disappeared altogether. The latter, in particular, is a periodic comet with a period of 61.2 years. At the time of its discovery in 1852, the comet increased in brightness regularly as it approached the sun and decreased as it receded. On its 1913 return the comet appeared relatively bright (about 7.5 magnitude) when first detected on September 26. It was scheduled to pass perihelion on November 26; in the two months prior to that date it should have kept increasing in brightness until it became easily visible to the naked eye. Instead, it kept on fading and eventually shrank to an object of the 17th magnitude barely discernible on a photographic plate taken on the night of November 22. It had become 10,000 times less luminous than it was at the time of recovery and henceforth was never seen again.

The study of brightness variations in comets is fraught with difficulties but can reveal some interesting facts. If comets were solid bodies shining by reflected sunlight, like the planets, their brightness would obviously depend only on the distance from the sun, or, to be precise, it would be inversely proportional to the square of the distance. Comets, however, do not just reflect the light of the sun. As we have seen before, they also emit some light of their own under the combined action of the solar wind and the often invisible solar radiation. Furthermore, unlike a solid planet, which always retains the same diameter, a comet develops a coma whose size depends on its distance from the sun, since the amount of particles contained in the coma increases upon nearing the sun. The total reflecting power therefore will also increase in proportion to the number of particles. Lastly, there are practical difficulties due to the fact that, as the comet travels along its orbit, not only does its distance from the sun change, but its distance from the earth varies as well. Thus the variations in brightness that we observe are different from the intrinsic ones.

To solve this problem we use the relationship

$$m = m_0 + 5 \log \rho + 2.5n \log r,$$

which relates the comet's distance from the sun (r) and from the earth (ρ) to the magnitude observed from earth (m) and to m_0, called the absolute magnitude of the comet. The value of m_0 is the apparent magnitude of a comet, as seen from earth, at a distance of 1 AU from both the sun and the earth, that is, when both ρ and r equal 1. The value of m_0 is characteristic for each comet. If all comets were identical, this value would always be the same, but observations have shown that it is not so. Thus we can say that the different values of m_0 show us to what degree comets differ from one another. The same is true also for n, a number that denotes how the brightness of a comet varies as a function of its distance from the sun. If the brightness changed only in accordance with the optical law of specular reflection, as is the case for planets, then n would equal 2. Since the values of ϱ and r can be derived from the orbit, and m is obtained from observations, the unknown factors are m_0 and n. These quantities depend on the physical state of the comet, so they can give us important information about it. Both m_0 and n can be easily derived from observations; we need only measure the apparent magnitude m from two different sightings and substitute the respective values in the equation to obtain two equations with the unknown quantities m_0 and n. In practice, we can obtain a more accurate result with the method of least squares to fit a large number of measurements of m.

The same formula, given r and m and $\varrho = 1$, enables us to calculate the magnitude of a comet as it would appear if we observed it at a distance of 1 AU. Using all the values of $m, m_1, m_2, m_3, \ldots, m_n$, obtained in successive sightings when the comet's distance from the sun was $r_1, r_2, r_3, \ldots, r_n$, respectively, we can determine its true variations in brightness as it approaches or recedes from the sun, just as if we were traveling in space with the comet at a constant distance of 1 AU.

The results are extremely interesting, and in figures 1.17 and 1.18 we show them for six comets. Plotted at the top of figure 1.17 are two normal cases: comet Arend-Roland (1957 III) and comet Bennett (1970 II). For both of these the distribution of individual points is tight enough to

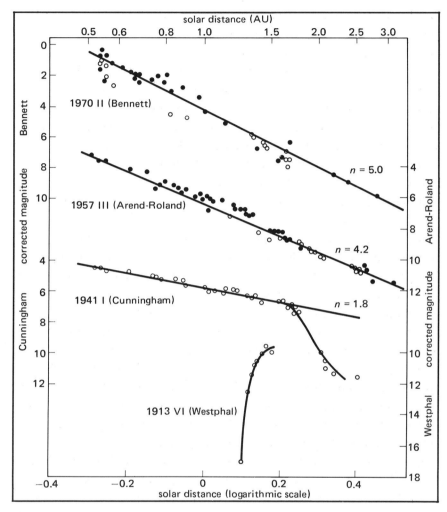

Figure 1.17 Brightness variations of four comets computed for a standard distance from earth of 1 AU and plotted as a function of the comet's distance from the sun in astronomical units. Magnitudes before perihelion are indicated with circles, and those after perihelion with dots. The luminosity decreases from top to bottom. (From *Sky and Telescope*.)

define a straight line whose slope depends on n. Note that in both examples n is greater than 2; so these comets reflected more light than a planet. Furthermore, the slopes of both lines remained constant throughout the observations; this shows that the law governing the brightness of the two comets remained the same during the entire period of their visibility. Quite different was the behavior of comet Cunningham (1941 I). Starting from a certain date, the comet's brightness increased very regularly with an exponent close to that of specular reflection, but in the brief period right after its discovery it changed to a much greater value of n. This is the reason why astronomers, using at first this large value of n, had predicted that its maximum luminosity would be very high.

Something similar also happened with comet Kohoutek (figure 1.18). However, in the case of this sly comet the value of n did not change abruptly but in a more gradual manner. Thus the points corresponding to observations before perihelion are not arranged along one or more straight lines but along a curve. In this curve the value of n at each instant is given by the slope of the tangent to the curve at that point. (In figure 1.18 it is given by the slope of the tangent at the point corresponding to the comet's solar distance at that instant.) In the few days around the first of the year 1974, the rate of increase in the comet's brilliance shot up abruptly, although nobody could see it during that period because of its proximity to the sun. When the comet reappeared, it was observed first by the astronauts on Skylab, and its brightness seemed to be decreasing according to a value of n just about as high as that which, to everybody's satisfaction, would have brought it to the maximum predicted brightness. In other words the comet did in fact increase and decrease greatly in luminosity, but only during those few days when it was closest to the sun and therefore invisible to us.

The variability of n corresponds to physical changes in the comet whose causes are as yet unknown. Recall that the variation of n does not correspond to the variation in brightness with the distance from the sun but to the variation of the law according to which the brightness changes. This means that the physical conditions of the nucleus change with changing n. As we have seen for comets Bennett and Arend-

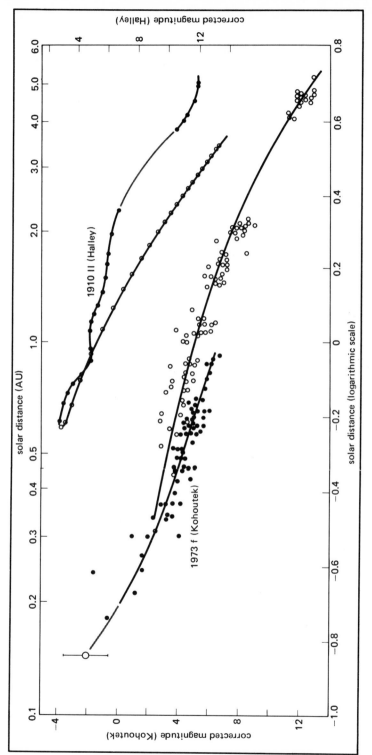

Figure 1.18 A comparison between comet Halley of 1910 and comet Kohoutek, before (circles) and after perihelion (dots). The graph must be read from right to left for the circles, and from left to right for the dots. (From *Sky and Telescope*.)

Roland, this phenomenon does not occur in all comets, nor does it always occur in the same fashion. For example, Halley's comet, which is periodic and perfectly regular, exhibited on its 1910 return an even stranger behavior (figure 1.18). While approaching the sun, its brightness increased regularly, almost along a straight line; while moving away, it underwent fluctuations corresponding to different and even negative values of n.

Yet this is not even the most extreme case. There are comets, such as Whipple-Fedtke-Tevzadze, that in some portions of their orbits show marked fluctuations in brightness (as high as 2 magnitudes) while in others they vary regularly according to an established and constant value of n. Still other comets have been observed to exhibit sudden flare-ups in brightness, and this is an even more mysterious and puzzling phenomenon. A typical case is that of comet Schwassmann-Wachmann 1, which has been studied very thoroughly. This comet, discovered by A. Schwassmann and A. Wachmann on November 15, 1927, moves in a nearly circular orbit between those of Jupiter and Saturn (figure 1.19) and is observable every year for a certain period of time. Owing to the position of its orbit, the comet's distances from the sun and from the earth do not change significantly. Accordingly, its brightness is not expected to fluctuate by more than 1.3 magnitudes and should always remain between magnitudes 18 and 19. Normally, therefore, the comet is visible only with the large telescopes, except sometimes, all of a sudden and for no apparent reason, it brightens by 5 or 6 magnitudes. From 1927 to 1950 thirty-two such outbursts were observed, and many others have taken place since then.

Comet Schwassmann-Wachmann 1 is not the only one that exhibits this kind of behavior. Similar phenomena had been previously observed in comet Holmes (1892 III), which returns every 7.3 years and was last seen in 1971. A more recent case is that of comet Tuttle-Giacobini-Kresàk of 1973. This comet returns every 5.5 years, but observing conditions are generally poor. It was first discovered by M. Tuttle in 1858 and, despite the shortness of its period, was not recovered until 1903 (by M. Giacobini). Attempts at observing it on eight subsequent returns were unsuccessful, and the comet was finally sighted by L. Kresàk in 1951.

Since then it was seen again in 1962 and 1973. In this last apparition the comet was expected to attain about the 13th magnitude at the time of the predicted perihelion passage on May 29. In effect, on May 20 it was slightly fainter than magnitude 14, but a week later, just before passing perihelion, it appeared ten thousandfold brighter, or about magnitude 4, and was thus visible even to the naked eye. Afterwards its brightness decreased, at first rapidly then more slowly, and by July 4 it was around magnitude 14. On July 6, when the comet had already been moving away from the sun for over a month, there was a new flare-up; the comet brightened to about the 6th magnitude and at the same time developed a brilliant nucleus, a short but well-defined tail, and, above all, a large luminous coma (figure 1.20).

No explanation has yet been found for this phenomenon. One view that it might be due to the comet's proximity to the sun, which activates all cometary processes, fails to explain how a flare-up could have occurred in comet Tuttle-Giacobini-Kresàk when it had already gone so far on its journey back to aphelion. Another seems to suggest that the flare-ups must be due to solar activity because they occur in comet Schwassmann-Wachmann 1, which always remains at more or less the same distance from the sun and thus in the same environmental conditions. If the latter supposition were the case, we would expect flare-ups to occur in most of the comets when, in fact, they take place only in a very few. Very likely both interpretations are correct, in the sense that the flare-ups are probably caused by sporadic solar phenomena that affect mainly a few exceptional comets whose peculiar structure makes them particularly susceptible to these processes. But the underlying causes of these flare-ups, as well as the mechanisms and processes involved, still remain a mystery.

END AND ORIGIN

DISINTEGRATION

The study of the physical structure of comets has shown that part of their material is blown away by solar wind and dispersed into interplanetary space. Nevertheless, they continue to exist and to appear as comets, since the lost material is continuously restored by matter expelled from the nucleus. In fact, gases and particles forming the head are wholly replaced every few hours, while the entire tail can be considered completely renewed within a few days. But the nucleus itself is not inexhaustible. Even though the loss of mass occurs only when the comet is near the sun, it is inevitable that sooner or later the whole comet will be exhausted.

According to some astronomers, this phenomenon may be so conspicuous that it can actually be observed in certain short-period comets which have been examined on many returns. One of the comets that best lends itself to the study of these processes is Encke's. Due to the shortness of its period (3.30 years), this comet, first discovered in 1786, has already passed perihelion 50 times (the last passage took place on April 28, 1974). In the mid-1960s the Russian astronomer S. K. Vsekhsvyatsky gathered all the observations made from 1786 to 1962 and found that during the intervening 176 years the absolute magnitude of the comet had decreased by about 5 magnitudes. On the basis of this result Whipple predicted that the decrease in brightness would continue to accelerate, and that the comet would vanish altogether by the end of this century. His predictions were later supported by Sekanina. Their validity, however, was questioned by other astronomers who reinterpreted the data taking into account other factors that had been overlooked at first—notably, the effect due to the variety of instruments used in the observations of the past two centuries. It was pointed out, moreover, that if comet Encke had undergone such a decrease in luminosity in the past 200 years, it should have been much brighter before 1700, and easily visible even to the naked eye a few centuries before then. A perusal of ancient chronicles, however, disclosed no evidence that such a comet had ever been observed. The controversy appeared to be resolved in 1974 with the publication of Kresàk's work. After reviewing all previous observations, the Czech astronomer found that comet Encke had indeed faded, but only by 1 magnitude per century, instead of 3 as Sekanina had suggested.

In conclusion, the dimming of a comet's light with the passage of time

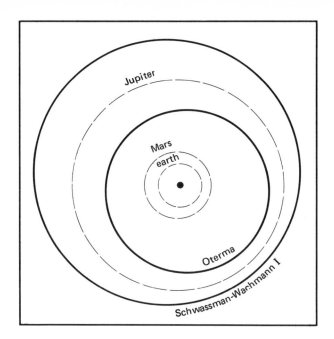

Figure 1.19 The orbit of comet Schwassmann-Wachmann 1. Also shown is the nearly circular orbit of comet Oterma (1925 II).

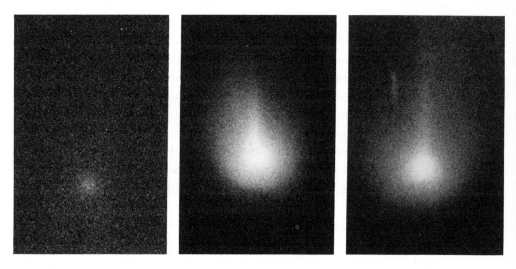

Figure 1.20 The flare of comet Tuttle-Giacobini-Kresàk observed at the Agassiz Station of the Harvard Observatory by R. E. McCrosky and C. Y. Shao. The first photograph was obtained on July 4, 1973, with an exposure time of 22 minutes; the second on July 8, 10 minutes; the third on July 9, 11 minutes. Notice that at some point between the first and second photograph the comet suddenly developed a brilliant nucleus, a coma, and a tail. (Courtesy of C. Y. Shao.)

seems to have been confirmed by observation, though a comet would not be extinguished in the course of a few centuries, but rather over several millennia. It is also obvious that the lifespan of a comet must be different in each case, since it depends on individual differences such as the amount and the type of the initial constituents, the length of time the comet spends near the sun, and, lastly, its perihelion distance.

SUN-GRAZING COMETS

It is easy to understand that the closer a comet comes to the sun, the stronger the sun's disruptive action will be and the shorter the comet's lifespan. This disruptive action is constantly at work but becomes fairly dramatic when some bold comets draw so close to the sun that they actually penetrate the upper layers of its atmosphere and graze its surface. Currently we know of eight such comets, all of which perhaps originated from a single object. Five appeared in the years 1668, 1843, 1880, 1882, and 1887 and were studied by H. Kreutz at the end of the nineteenth century; the other three were observed in 1945, 1963, and 1965. The most recent of these daring comets is the beautiful comet Ikeya-Seki, named after the two Japanese amateur astronomers who discovered it on September 18, 1965. It first appeared in the evening sky at the end of October, then it grazed the sun, penetrating deeply into the corona and passing 464,000 km from its surface. When the comet reappeared, its nucleus had split in two. After studying the orbits of the eight comets, Marsden was able to prove that comet Ikeya-Seki and that of 1882 were once a single object, which can perhaps be further identified with the comet of 1106.

Here then is an example of the continuous and progressive breakup of a comet resulting in its total destruction. Thousands of years ago, there may have existed a single comet that gave birth to some of these eight or to an intermediate generation of parent comets, including that of 1106. From this last one, comets 1882 II and 1965 VIII were born, and in 1965 the nucleus of the latter split in two, from which, in turn, two other comets of different periods were born. One of them will return in 878 years, and the other, apparently, in no less than 1,055 years.

COMETS BIELA AND WEST

A much more radical process of disintegration was observed last century in a comet that did not graze the sun—comet Biela.

The history of this comet is quite interesting. Baron W. von Biela, a captain in the Austrian army who was stationed in Bohemia, discovered it on the night of February 27, 1826. It was later identified with the comets of 1772 and 1805 and turned out to be a periodic comet revolving about the sun in an elliptical orbit with a 6.7-year period. Subsequently, it was duly observed on its 1832 return but could not be seen in 1839 because of unfavorable observing conditions. It was detected again on November 25, 1845, when it once more approached the earth and sun. Quite likely, the comet had pursued this path for hundreds of years, and during the first month and a half of observations there was not a hint of the catastrophe that was about to take place. At the beginning of January 1846, the comet brightened to a marked extent. Then on the night of January 13, to the great amazement of astronomers, there appeared a second faint comet right next to the main one. By the following night two parallel tails had formed a few arc minutes in length. Afterward, the two comets brightened and faded repeatedly, not in phase but alternating, so that first one body would appear brighter and then the other. Toward the end of February, despite the fact that the two heads were already more than 250,000 km apart, there developed a third tail that seemed to link them like a bridge. By the end of April both comets had vanished.

The two comets reappeared in 1852. On the evening of August 26, the Jesuit Secchi sighted the main comet at a considerable distance from its expected position, but he could not detect the second one until September 15. The distance between the two comets had increased to 2,800,000 km. The appearance and brightness of the two heads kept changing from one night to the next, and at the end of September they vanished from sight. Neither comet has been seen since. Astronomers searched for them at the times of their appointed returns in 1859, 1866, and 1872 but without any success. Comet Biela had been all but forgotten, when on the evening of November 27, 1877, it reappeared once again, though in a strange fashion. That night, as the earth was passing the point in its orbit closest to the comet's path, a fantastic meteor shower fell from the sky. For more than six hours the sky was streaked by thousands of luminous trails, all of them apparently radiating from a point near the star Gamma Andromedae. It was estimated that at least 160,000 shooting stars must have appeared in that remarkable display of fireworks. Evidently, the comet had broken up into the meteoritic fragments that until a few decades earlier had remained united in a single compact block but now had wandered apart and were strewn along the comet's orbit. These meteors, however, could not have come directly from the comet, which according to computations should have passed that point about three months earlier, but from a trail it had left behind on its path.

On November 27, 1885, there was another meteor shower that confirmed the interpretation of the previous one, while the absence of showers around the same date in the years between 1877 and 1885 showed that the meteors were not distributed along the entire orbit. Some of these meteors, called Andromedids, still fall to earth, but no longer with the same profusion or concentration. Every so often, they still appear between November 15 and December 6, thus proving that the disintegration of the main body has been followed by a slow dispersion in space.

The celestial history of comet Biela ends here, but its terrestrial story goes on. A few years ago Marsden suggested the possibility that the comet might not have been completely destroyed. In his view the 1846 disruption caused the nucleus to split into two unequal parts: one consisting of the surface material, rich in ice and meteoritic dust; the other of the more solid central core. Accordingly, the first would have produced a flashy but ephemeral comet, since vanished, while the second would still exist as a dark body similar to an asteroid. Assuming that the nongravitational forces ceased to act in 1859, when the comet was last seen, Marsden computed an orbit for the remaining (and most important) fragment and found that it would pass perihelion on December 21, 1971, at a distance of only 0.051 AU from the earth. The search for the fragment was not successful, but in the course of these observations some rather faint asteroids were detected, although their connection with the remnants of comet Biela was not quite clear. Curiously, it was during an extensive search for this body with the Schmidt telescope of the Hamburg Observatory that astronomer L. Kohoutek discovered in April 1973 the comet that bears his name and would later become so notorious.

The disruption of comet Biela, although perhaps the most dramatic, is certainly not the only example of the breakup of a cometary nucleus; we know in fact of a dozen comets with two or more nuclei, and not all of them are in the sun-grazing group.

The most thoroughly studied case is the very recent one of comet West. This comet was discovered in November 1975 on plates obtained with the Schmidt telescope of the European Southern Observatory at La Silla, Chile. It passed perihelion on February 25, 1976, and at the beginning of March it appeared in the dawn sky even brighter than predicted —a truly lovely sight. Both professional and amateur astronomers had all sorts of instruments trained on it, but its head and nucleus could be observed with the naked eye. Around the time it passed perihelion at a distance of only 7° from the sun, the comet's head was so brilliant that it could be seen in full daylight. Observations began in early March, and on the morning of the 5th it was first noted that the nucleus was no longer single and diffuse, as it had been on the 3rd, but had split in two. Observations of March 8 revealed three nuclei, which became four after March 11.

In the past only two comets had been observed to break up into more than four nuclei: the "September comet" of 1882 and the periodic comet Brooks on its 1889 return. The evolution of the four nuclei of comet West was photographed, leaving us the first accurate record of the displacements and variations in brightness of the comet's various components during the early phases of the catastrophe. Sekanina used these observations to compute the apparent paths of the three nuclei, labeled $B, C, D,$ with respect to the reference nucleus A in figure 1.21a, and to trace the various components back to the time they had begun to separate. His calculations showed that the first to separate from A was nucleus D on February 13, then B on February 22, and lastly C on March 5. According to Sekanina, it is possible that B and D separated from A as a

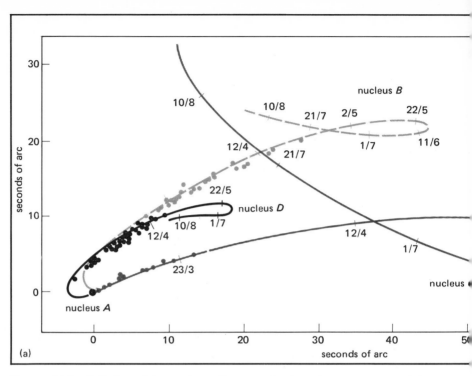

Figure 1.21 Splitting of the nucleus of comet West into four separate nuclei (March to April 1976): (a) the positions of three of the nuclei with respect to nucleus A in the days after the fragmentation (the segments without dots indicate the expected positions for the subsequent months); (b) the estimates of the visual magnitudes of the four nuclei made by V. T. Helms with a 40-cm telescope.

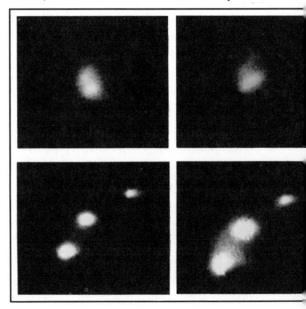

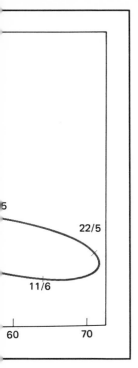

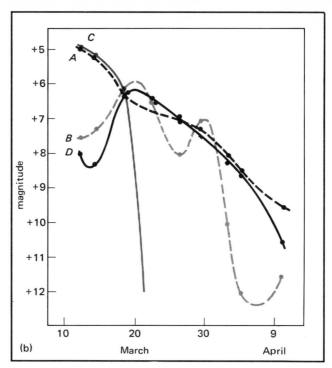

5

22/5

11/6

60   70

+5
+6
+7
+8
+9
+10
+11
+12

C
A
B
D

magnitude

10   20   30   9

(b)

March   April

Below, photographs of the nuclei obtained by S. Murrell and C. Knuckles with the 61-cm telescope at the New Mexico State University in Las Cruces: *top row,* the set of plates obtained in yellow-green light from March 8 to March 24; *bottom row,* the set obtained from March 31 to April 8 with an image intensifier sensitive to red. (New Mexico State University Observatory.)

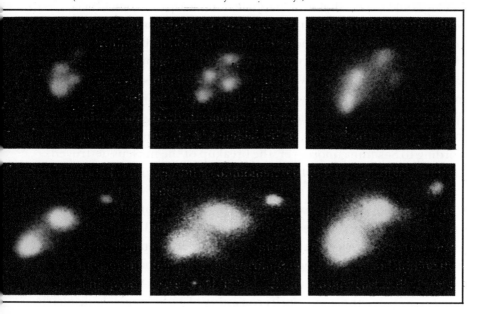

single nucleus and that only later B separated from D; the latter breakup would have taken place on February 26, soon after perihelion passage.

The four nuclei appeared to have different magnitudes, and on the following days did not retain the same relative brightness. Rather, nucleus C, which appeared as bright as nucleus A (brightest of the four) when it was first sighted, faded faster than all the others and was last seen, very faint, on March 27. At the same time both D and B were brightening rapidly, until they equaled and even exceeded nucleus A. After March 22, while D and A were slowly becoming fainter, nucleus B twice underwent marked upward and downward fluctuations in brightness, in a seemingly capricious fashion (figure 1.21b).

The photographs documenting the variations in brightness are shown in figure 1.21. Most striking in these pictures are the faint smudges that can barely be seen trailing from each new nucleus in the opposite direction from the sun. They are particularly noticeable in the photographs taken during the first ten days in April, when the nuclei were already fairly well separated. These smudges are embryonic tails, and we are actually witnessing the birth of new comets from newly formed nuclei.

One could expect that what happened in the case of comet Biela should also occur for comet West, that on its next return we should see at least three objects. This is not at all certain, however, because the nuclei might draw farther and farther apart, and the new comets would then return at different times. What is certain in any case is that neither we nor our close descendants will ever be able to verify this point. Comet West describes an orbit so large that even if it were to come back to the neighborhood of the earth, it would not be for a few million years. According to Sekanina, moreover, the breakup of the nucleus destroyed the comet's brightness, and none of the new comets will ever be as spectacular as their parent was in 1976.

Why a comet should break up when revolving about the sun at a distance that could be considered safe, may be explained by the fact that its stability depends to a great extent on the mass and composition of the nucleus. With respect to the latter, we should recall that there might be two types of nuclei, as was suggested by Marsden and Sekanina. All other conditions being the same, nuclei consisting of a solid nonvolatile matrix will be much more cohesive than those consisting mainly of ice loosely interspersed with small solid particles. The two types of comets will also have different endings. After losing all volatile materials, the solid-matrix nuclei will evolve into inert objects like the asteroids. But in comets of the second type the evaporation of volatile substances will cause an increasing number of meteoritic particles to be scattered along their orbits; these swarms of particles will continue to wander in interplanetary space until the day when, yielding to a planet's attraction, they will penetrate into its atmosphere and in the process of dissolution give birth to the short-lived, luminous wakes of shooting stars. Thus even the last particles of the comet will shine in the sky.

VIOLENT END

The end of a comet can occur not only in the tame way we have described but also in very frightening circumstances for the inhabitants of earth.

On June 30, 1908, at 7 o'clock in the morning, a giant meteor fell in a deserted region of Siberia, at 60°55′ latitude north and 101°57′ longitude east (figures 1.22 and 1.23). From the region where it fell the bolide came to be known as the "Tunguska meteorite." Even though it was full daylight, people who lived near the path of the trajectory saw a fireball cross the sky in a few seconds from southeast to northwest, leaving behind a thick dust trail that appeared from directly below its path as a huge pillar etched against the sky. Near the end of its path the meteor exploded with a booming sound that was heard more than 1,000 km away. The shock wave had astonishing effects within a radius of tens of kilometers. At the Vanavara farmhouse, 60 km from the explosion, a man sitting on a porch experienced a strong flash of heat and was thrown several meters, losing consciousness. "I was shaken by a tremendous explosion," he said later. "Every building on the farm shook, glass panes shattered and window frames were jolted out of place." At the same time flames and clouds of smoke rose from the site of the explosion.

The tremors caused by the impact were recorded by several seismographs located at great distances from the site, and the aerial shock wave was registered on the microbarographs of the Potsdam Observatory in

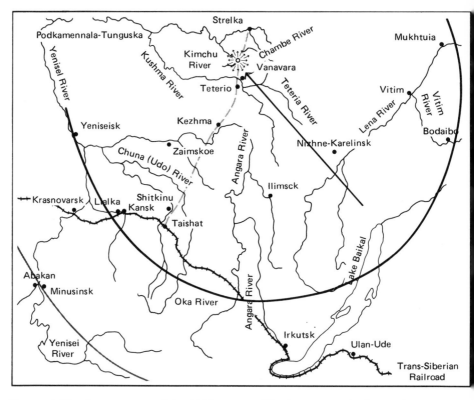

Figure 1.22   Impact zone of the 1908 meteor. The long arrow indicates the meteor's trajectory. The dark curved line marks the boundary of the area of visible phenomena. The light gray line is the boundary of the area of audible phenomena. The dashed line is the expedition's route. (From E. Krinov, in *The Solar System*, vol. 4.)

Figure 1.23 *Above,* region at 8 km from the point of impact. Almost all the trees were blown down in the same direction. *Below,* the presumed point of impact in a picture taken during the 1927 scientific expedition led by A. Kulik. The lack of a crater is quite evident. There is only a slight depression about 7 km in diameter. The central plane is totally devoid of trees. (Courtesy of E. Krinov.)

the suburbs of Berlin. The following night the skies over Siberia and Europe were unusually bright. Even at low latitudes, for instance in the Caucasus, there was enough diffuse light around midnight to read the newspaper. The phenomenon persisted for several nights, although gradually decreasing in intensity, and then stopped altogether. On the basis of measurements taken by C. G. Abbot in California in 1908, years later the astronomer V. G. Fesenkov was able to determine that the transparency of the earth's atmosphere had been considerably below normal from the middle of July to about the end of August 1908. In Fesenkov's view, this proved that an enormous amount of pulverized material had been released into the atmosphere at the time of the meteor's fall.

The Tunguska region was sparsely populated and not easily accessible. These facts, coupled with a certain reticence on the part of the local inhabitants caused by superstitious fears of the phenomenon, made it very difficult to locate, let alone reach, the exact place of the meteorite's fall. It was not until 1927 that a scientific expedition led by L. A. Kulik was finally able to reach it.

The results of this expedition proved of the greatest interest. Thousands of trees were found to have been felled and scorched over a radius of more than 30 km from the center of the explosion, which was clearly indicated by the direction of the trees themselves all converging toward it. Upon reaching the central area, the expedition found a slight depression about 7 km in diameter, which was assumed to be the main crater (figure 1.24). Northeast and northwest of the depression, the explorers discovered a great number of deep holes with diameters ranging from a few meters to several tens of meters, which in Kulik's view were meteorite craters formed by fragments of the main body. After spending only two weeks on the site, the expedition was forced by a shortage of provisions to return to Leningrad. Kulik, however, set out immediately to organize further expeditions whose main goal was to search for meteoritic material both at the bottom of the minor craters and in the epicentral zone of the tundra.

The combined efforts of all these expeditions led to several discoveries, foremost among them one that nobody had anticipated:

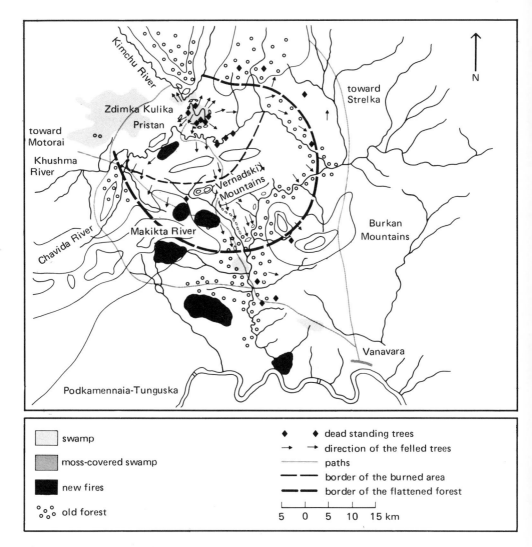

Figure 1.24   Large scale map of the Vanavara area. This is the center of the phenomena associated with the impact of the giant meteor which fell in Siberia on June 30, 1908. (From E. Krinov, in *The Solar System*, vol. 4.)

neither Kulik nor anybody else ever found a trace of large meteoritic fragments or of the main body. The craters discovered by the first expedition turned out to be natural local features, and none of them was found to contain any meteoritic material. Evidently, either the meteor had been pulverized by an explosion while still in the air, or it had never existed as such. With the aid of aerial reconnaissance, surveys of the area where the mysterious body had fallen revealed that the depression was not circular in shape but elliptical, with the major axis of the ellipse oriented in the direction of the trajectory. In addition, a great deal of fine meteoritic material was found scattered over a very large area as far as 60 to 80 km to the northwest.

As a result a new hypothesis began to gain wide acceptance, the very same one that had already been independently proposed by F. L. Whipple and I. S. Astapovich around 1930. A new expedition led by geochemist K. P. Florensky set out in 1962 to gather samples of the meteoritic dust and establish its origin, and thereby confirm or disprove the novel idea. The findings of this expedition appeared to support the hypothesis fully, and thus confirm a most extraordinary fact: on June 30, 1908, the earth had been struck by a comet.

Everything could now be satisfactorily explained. The fine material scattered to the northwest of the epicenter corresponded to the coma, which, as we know, forms mostly away from the sun. Both the absence of large fragments and the breakup prior to impact were quite consistent with the loose and low-density structure of comets of the dirty snowball type. Furthermore, the white nights and the reduced transparency of the earth's atmosphere in the period after the fall could now be easily explained by assuming that significant amounts of dust and gases had been associated with the body, which is in fact the case for comets.

But is it possible that even a modest-sized comet could steal so close to earth without being detected and suddenly fall on it?

It is entirely possible. Let us assume that the comet's orbit is such that the last segment before perihelion (when the comet is brightest being closest to the sun) more or less coincides with the direction of the sun. In this case the comet is present in the daytime sky in close proximity to the sun and cannot be detected. Several such cases are known: comet Mrkos

(1957 d), for instance, was discovered by many with the unaided eye when it was already at perihelion, while many other comets were first detected during total eclipses when they were already near the sun and at the height of their splendor.

This is probably what happened in the case of the "Tunguska meteorite." Since it fell from the southeast at 7 o'clock in the morning, it approached the earth almost exactly from the direction of the sun and therefore must have remained invisible to the very last.

After all, then, there is still some reason to be frightened of comets: not when they sweep across the sky in regal splendor, or flit by like pale ghosts, but rather should they approach us unseen, their small bodies lost in the overwhelming radiance of the sun, and suddenly crash on earth sowing death and destruction. It happened once, and it can happen again. If the 1908 comet had fallen just four and three quarter hours later, owing to the earth's rotation it would have hit the city of St. Petersburg, which is at about the same latitude as the actual point of impact.[3] It must be pointed out, however, that collision with a comet is an extremely rare event and that the surface of the earth is covered for the most part by oceans and deserts. Therefore the probability that a comet will hit a populated area becomes very small.

ORIGIN

In the foregoing pages we have seen that orbital variations profoundly affect our ability to observe periodic comets. They might lead us to believe that a new comet has been born when in reality we are seeing an old one deflected into a smaller orbit and coming to perihelion closer to the sun, or, conversely, that a comet has vanished when planetary perturbations have simply widened its orbit. We have also seen how comets can undergo actual transformation and disintegration for physical reasons and appear as new bodies following the disruption of an old one, how comets may be destroyed by the heat of the sun or by their own structural weaknesses and even disappear abruptly when they could be logically expected to increase in brightness. All these facts are so many

3. Modern Leningrad.—Trans.

stumbling blocks along the astronomer's path as he attempts to solve the mystery of their end and, above all, their origin.

Apart from all this, we are now convinced that comets not only have an end, like everything else in the universe, but also a relatively short life-span. Consequently, assuming that all comets originated at a given epoch, they should all have disappeared after a certain period of time that is not too long by cosmic standards (a few million years, for example). Of course, we could assume that their lives are unfolding just now, when we can observe them; but this interpretation, though possible, is too anthropocentric. If anything, a much more reasonable assumption would be that comets originated at the beginning of the solar system, when the sun, planets, and satellites were all evolving from a primordial nebula.

Several theories have been formulated to explain the origin of comets, but in practice they can be reduced to three: (1) comets continuously originate from interstellar space and become more or less bound to the sun after being captured by the massive planets; (2) comets formed and continue to form in the solar system; (3) comets originated within the solar system in the remote past.

An interstellar origin has often been advocated. According to theories formulated in the last century, comets formed from condensations in a cloud of interstellar matter when the solar system happened to pass through it. A more recent version of this theory, proposed by R. A. Lyttleton about twenty years ago, envisages instead a process of accretion from a homogeneous cloud of interstellar matter due to perturbations caused by the sun's passage. Both theories are in disagreement with our current knowledge. Lyttleton's, in particular, would result in the formation of nuclei of the gravel-bank type; but in the last few years, as we have seen, both the consensus of opinion and, more importantly, increasing observational evidence have favored instead the dirty snowball model.

The existence of parabolic and hyperbolic orbits, on the other hand, would seem to require an interstellar origin. As already mentioned, however, parabolic orbits correspond to an extreme case and quite possibly do not exist. As for hyperbolic orbits, it must be pointed out that we are observing orbits that have already been altered by planetary perturbations and that we must therefore trace them back to the original ones. In this context it is significant to note that a few years ago some astronomers reconstructed the orbits of several hyperbolic comets, taking into account the perturbing effects of the great planets. They found that the initial orbits were in reality greatly extended ellipses.

Following this new approach only two out of nineteen comets still appeared to retain hyperbolic orbits, although in both cases the differences with parabolic orbits were so small as to be well within the margins of error.

In conclusion, it seems correct to assume that, as a rule, cometary orbits at first are ellipses, although so elongated that their aphelia may fall well beyond Pluto's orbit. Therefore, except for very special cases, comets do not appear to originate from interstellar space.

Let us consider now the other two alternatives. The theory that comets originate within the solar system was often proposed in the nineteenth century. Currently its most fervent advocate is the Russian astronomer S. K. Vsekhsvyatsky. In his view, short-period comets formed, and are still forming, from volcanic eruptions on the great planets or on their satellites. Naturally these eruptions must be violent enough to impart to the ejected material a speed greater than escape velocity, which is that value of velocity a body must have in order not to fall back on the planet.

It has been computed that the initial velocities required to eject these "proto-comets" from Jupiter and its satellites are of the order of 5 to 7 and 1 to 3 km/s, respectively. According to Vsekhsvyatsky, his own theory is confirmed by the distribution of the elements of cometary orbits, by the similarity between the chemical composition of the comets and that of the atmospheres of the great planets and their satellites, and lastly by actual observation of eruptions on the planets, such as the white spot detected on the surface of Saturn in 1933. Then again, many comets of Jupiter's family have shown close orbital approaches to the planet just prior to discovery, which would be proof of their being ejected practically under the eye of the observer. According to calculations made by the Russian astronomer, at least three others originated instead from the region of Jupiter's satellites. With regard to the high energies involved in these processes, he points out that 10^{26} and 10^{27} erg, respectively, were expended in the eruptions of Krakatoa (1883) and Tomboro (1815), while the energies involved in the cataclysmic eruptions of the tertiary and quaternary periods must have been of the order of 10^{29} and 10^{30} erg. The energy necessary to eject a comet with a mass of 10^{13} to 10^{15} g from one of Jupiter's satellites has been estimated to be between 10^{25} and 10^{27} erg, which is quite comparable with that of a volcanic eruption on earth.

To the extent that parabolic comets have the same physical structure as periodic ones, their volcanic origin could also be confirmed by the computation of the energy necessary to eject a given mass. Unlike periodic comets, however, they would have originated in a more remote past when ejection velocities from the parent planets were considerably greater, but in any case no earlier than 300 million years ago.

Independently from the theory of a volcanic origin, a number of astronomers have envisaged the existence of a sort of cometary belt in that portion of the solar system bound by the orbits of Jupiter and Saturn. This question is still quite controversial and will not be discussed here for two reasons: in the first place, this hypothesis postulates the existence of the belt without giving any physical reasons for it; second, it has been proved that comets originating in this region could never attain long periods. Consequently, it would be necessary to adopt two different theories: one to explain the origin of short-period comets and the other to explain the origin of comets with long periods or quasi-parabolic orbits.

Let us consider now the third and last theory, formulated and developed by J. H. Oort from 1950 on. By examining the distribution of the semi-major axes of about fifty selected comets that passed perihelion between 1850 and 1952, Oort discovered that all their aphelia lay at a distance of about 50,000 AU, that is, at distances greater than 30,000 but smaller than 100,000 AU, with maximum concentration around 50,000 AU. This result proved that the comets had originated from a spherical shell centered on the sun, whose outer surface had a radius no greater than 100,000 AU (figure 1.25). Considering that the nearest star (Alpha Centauri) is at a distance of 250,000 AU, we can see that this cometary zone extends about halfway between the sun and the nearest star. Even

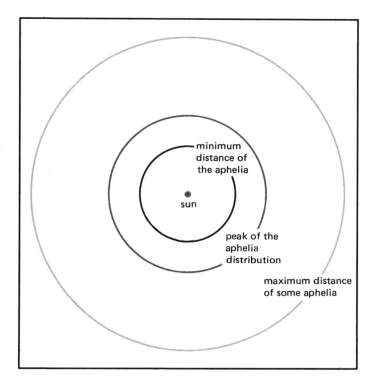

Figure 1.25   The "reservoir of comets" according to J. Oort's theory. The aphelia of the cometary orbits chosen by Oort fall in an annulus with an inner radius of about 30,000 AU and an outer radius of about 100,000 AU. Most of the aphelia are located in a spherical shell at about 50,000 AU. The real distribution is three-dimensional, not flat; therefore this illustration should be considered as a planar section through the sun (which is not in scale). In this scale Pluto's orbit would appear as a circle less than 1/10 mm in diameter. Alpha Centauri is at a distance of 2.5 times the maximum radius.

at this enormous distance, however, the gravitational attraction of the sun is still strong enough to hold celestial bodies in its power.

Within this shell comets hover around the sun, barely moving. But due to their position they may be perturbed by the passage of nearby stars, which can have two opposite effects: one is to draw comets away from the gravitational field of the sun, the other is to deflect them toward the inner part of the solar system. According to Oort, only one comet out of 100,000 would follow the second course. On the basis of this percentage, and the frequency of appearance of these comets, he concludes that the number of comets currently stored at the fringes of the solar system must be of the order of hundreds of billions.

In the more favorable case the comet begins to move toward the sun, and provided it approaches within a relatively short distance, it can finally be seen from earth. A comet of this type is identified by computing its unperturbed orbit. If it has never been observed before, it is regarded as a "new" comet. If it chances to pass close to one of the great planets, the comet may be captured and deflected into a very small elliptical orbit; otherwise it will recede along its very extended orbit and will not return for thousands of years.

At this point it is interesting to recall that when a comet nears the sun gases and dust are released from its head. The gases are revealed by the fast developing straight tails and, spectroscopically, by their band emission spectrum. The presence of dust is shown by curved tails and by a continuous spectrum of reflected sunlight. It has been shown that, after several close encounters with the sun, dust particles are dissipated more rapidly than gases. As a result the continuum must be very intense when a comet makes its first appearance. Subsequently, gas emission bands become ever more prominent above the weakening continuum, which ultimately disappears. This difference becomes quite evident if we compare the spectrum of an "old" short-period comet like Encke's (figure 1.12) with that of a new comet such as Arend-Roland (1956 k), and thus it provides us with an important additional criterion to distinguish the two types of comets.

This theory appears well founded and supported by observational evidence, but it does not explain how this enormous reservoir of comets could have formed at such a great distance from the sun. Oort suggests that it might be a remnant of the primordial nebula from which the solar system originated. The reservoir need not have formed so far away; it could have been born in the same region as the planets and later driven outward and dispersed by planetary perturbations. In this model comets would have formed as minor condensations during the formation of the major planets or in the same as yet unexplained event that produced the asteroid belt. These primitive fragments had unstable orbits: some of them fell on the planets and some were thrown by strong perturbations toward the fringes of the solar system. Thus the bulk of the comets would have formed five billion years ago and would have been transferred to the outlying reservoir in the following half a billion years. Many of them were later freed by stellar perturbations and flung into interstellar space; many others remained instead in the swarm and are still held there in the form of solid blocks of frozen gases. For this reason the reservoir has been jokingly called a deep-freeze storage of comets. When one of these bodies is accelerated toward the sun, its age-long hibernation comes to an end. The comet approaches ever closer and finally ignites.

According to a theory formulated by K. A. Shteins between 1961 and 1967 and currently under study at the Institute for Theoretical Astronomy in Leningrad, the capture process does not appear to take place in a straightforward way, but in several stages. The initial phase consists of a swarm of comets of nearly parabolic orbits located at the periphery of the sun's sphere of influence (as suggested by Oort). The second phase is that of stellar perturbations, which are not believed to alter the semi-major axes of the orbits to any significant extent but rather push their perihelia toward regions closer to the sun. In the third phase a comet begins to be affected by planetary perturbations, whose main effect is to shorten the semi-major axis and change its quasi-parabolic orbit first into an elliptical orbit and then into a nearly circular one. During this phase the orbital inclination keeps on decreasing, and the comet's orbital plane moves ever closer to the plane of the ecliptic, where all the planets revolve. This new orbit takes the slow moving comet very close to the paths of the massive outer planets. As a result of these long-lasting approaches, the planets tend to capture the comet and further change its orbit, deflecting it more and more toward the inner regions of the solar system and ultimately into Jupiter's family. Thus little by little a number of comets move from the reservoir toward the center of the solar system, and, as their orbits become increasingly smaller, they find themselves more or less bound to these inner regions and so become visible to us.

This theory shows comets under an entirely new light: far from being monstrous harbingers of ill fortune to a small humanity of no consequence to them, they appear to be the remnants of that matter which in a most remote past gave birth to the sun, the planets, and eventually to man himself. Although a great deal of that material was lost in interstellar space, after a wait of many long years a large fragment does occasionally come back to its origin. It returns as a comet which, after revolving for a long time about the sun, will ultimately dissolve in a nebulosity that seems to vanish into nothingness—a tenuous nebulosity similar to the primordial cloud from which the comet was born a long time ago.

2   PHANTOMS OF THE SOLAR SYSTEM

NATURAL EARTH SATELLITES

For those of us who dream of life on another planet, one of the most evocative fantasies is that of a night lit by many moons. Imagine them of various sizes and at different distances, so that some will look larger and some smaller. Picture them as they travel across the sky, waxing and waning and eclipsing one another; think of the nights when several of them have risen above the horizon, forming an ever-changing pattern of light and shadow and giving the landscape the appearance of an enormous stage lit by many floodlights. If you can capture all this in your mind's eye, even the loveliest moonlit night on earth, which after all has been endowed with only one companion, pales in comparison and seems so much the poorer. Our feeling of regret is certainly lessened by the realization that planets like Mercury and Venus have no satellites at all, while others, Mars, for example, have several but every one of them so small that we can be grateful for our one moon. And even when a planet has many satellites (Jupiter, for instance, has thirteen), it is not certain that they all put on a spectacular show. Some may be so small or so far away that at best they look like faint stars, while at worst they might be completely invisible to the naked eye. From this it is but a short step to the thought that the moon might not be the earth's only satellite but perhaps just the brightest. As a matter of fact a number of observations seem to point to the existence of other natural satellites of earth of different types and origin, both large and small, permanent and transitory.

SATELLITE CLOUDS
Satellite clouds have been thoroughly investigated by the Polish astronomer K. Kordylewsky, who for many years searched for earth satellites at two special points on the moon's orbit, the so-called Lagrangian points, L_4 and L_5.

It all started from a theoretical prediction of a more general nature that found confirmation in observational evidence only a century later.

G. L. Lagrange, the famous French-Italian mathematician, published in 1772 a paper examining the motion of three bodies of known mass and subject only to their mutual attraction. In it, he stated that, when three bodies occupy the points defining an equilateral triangle, their motions must be such that the triangle will remain equilateral. Let us assume, for example, that two of the bodies are the sun and a planet; should there be a third body on the planet's orbit at a distance of 60°, while revolving about the sun, it would always have to keep the same distance (60°) from the planet (figure 2.1). The same condition can of course be satisfied by a fourth body, provided its position is symmetrical to the third with respect to the first two. It was established later on that the three bodies need not be located precisely at the vertices of an equilateral triangle: once the positions of the first two are established, it is sufficient for the third one to lie in the vicinity of the third vertex, about which it will make small amplitude oscillations. In a four-body system, the two free vertices that are symmetrical with respect to the line joining the other two bodies are called Lagrange's libration points and are termed L_4 and L_5.

For many years this case was regarded as a simple mathematical curiosity. Then in 1906 a strange little planet whose apparent motion was unusually slow was discovered at the Heidelberg Observatory. This suggested a greater distance than that generally attributed to the other asteroids, which are all situated between the orbits of Mars and Jupiter. Calculations showed that its orbit was quite close to that of Jupiter and that it preceded the planet by 60°. Thus the new body, together with Jupiter and the sun, fully satisfied the conditions of the problem set by Lagrange, for which it provided the first concrete example. At the end of the same year another small planet was discovered near the second Lagrangian point in the sun-Jupiter system. Since then, more and more asteroids have been discovered near the two Lagrangian points, and at present twenty-two are known (figure 2.1). The asteroids of the two groups have been named after the heroes of the Trojan war, and are collectively known as Trojans. The group preceding Jupiter on its circumsolar path is that of the Greeks, while the group following the planet is that of the Trojans. Only two small planets do not conform to this mythical ordering, Hector and Patroklos: the former is in the Greek group and the latter in the Trojan.

Several years ago astronomers began to investigate the possibility that

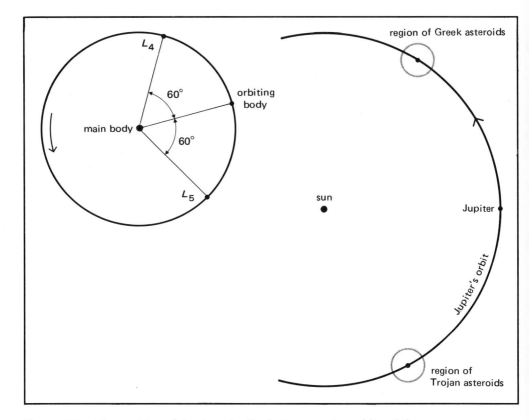

Figure 2.1   The position of the three bodies in Lagrange's problem: *left*, a general case; *right*, the positions of the Greek and Trojan asteroids at the L_4 and L_5 points of the Jupiter-sun system.

asteroids might also be clustered around the Lagrangian points in the earth-moon system. Naturally, should these bodies exist, they would be earth satellites just like the moon. Around 1951 K. Kordylewsky of the Krakow Observatory undertook a telescopic search at the predicted locations, looking for rather faint, starlike objects that move among the stars with the same apparent velocity as the moon. After years of patient observations he reached the conclusion that no object brighter than the 12th magnitude exhibited the required properties. He was about to give up the search when one of his colleagues suggested that perhaps there was no single body, big or small, in either of the two regions but rather a whole swarm of smaller objects and particles that by reflecting sunlight could produce a zone of diffuse luminosity, somewhat like a milky spot on the sky. Perhaps this spot could be visible even to the unaided eye, but it would have to be extremely faint to have escaped detection in the past.

The new search presented serious difficulties, and just to make it feasible several conditions had to be met. In the first place, since the hypothetical clouds were believed to shine by reflection, they would attain maximum luminosity when they lie in opposition to the sun (in a position analogous to that of the full moon). At such times the moon is always more than half full, and therefore it would have to be well below the horizon to make observations possible. On the other hand, the Lagrangian point where the luminous spot was to be found would have to be high enough on the local horizon to minimize the effects of absorption and scattered-light background introduced by the earth's atmosphere. Another requirement was that the position of the clouds in the sky should not coincide with either the Gegenschein or the Milky Way.[1] Finally, observations would have to be carried out with very good visibility, far from city lights and possibly high on the mountains.

In order to satisfy the last requirement, Kordylewsky decided to carry out his observations at the high altitude Czech observatories on the Lomnica Mountains, at Kasprowy Wierck and Skalnate Pleso. Starting in October 1956, he was able to detect visually two luminous patches corre-

1. Gegenschein is a faint luminous patch that can be seen on the sky in the region directly opposite the sun. The phenomenon is associated with interplanetary matter.

sponding to the points L_4 and L_5. They were very faint, much fainter than the central region of the Gegenschein, yet in the few instances when they could be observed for two consecutive nights, they appeared to move on the sky with the same apparent motion as the moon. Discovery seemed assured but could not be completely certain on the basis of observations carried out at the threshold of visibility. Consequently Kordylewsky set out to obtain more convincing photographic evidence. Finally, on March 6 and April 6, 1961, he was able to obtain plates of the cloud in the region of L_5 in which two spots could be seen at a distance of about 8° from each other. In September of the same year, on two different nights, he succeeded in recording photographically two other spots adjacent to each other and each of 5° diameter, but this time in the region of L_4.

One might imagine that as a result of these findings no doubts would remain about the fact that the earth has not just one or two satellites besides the moon but two whole clouds of extremely small, but very numerous, satellites. In the euphoria that followed the announcement, some astronomers began to make plans for further study of the clouds while others went so far as to suggest that the largest fragments could be used for space travel. But even photographs, when taken under unfavorable conditions, are as open to criticism as visual observations, and many astronomers received Kordylewsky's result with a certain degree of skepticism.

The Polish astronomer, however, was not the only one who believed in the existence of the two clouds of earth satellites. One of them was also seen by two Americans, J. W. Simpson and R. G. Miller, who in January of 1964 observed it twice in the same week at the Locksley Observatory on the Santa Cruz mountains in California. They have since claimed to have seen both clouds several times. Moreover, they have collected testimonials from a number of visitors who were able to spot them on the basis of a summary description of the phenomenon but did not know in which regions of the sky they should look for them.

Nonetheless, uncertainty still remains, owing mostly to the doubts of all those who have not yet been able to see them. Consequently, the two clouds of earth satellites, believed to accompany the moon 60° ahead of

and behind its journey, remain to date rather fictitious. But the search
for new evidence is open to everybody, including the layman who does
not have any instruments; as we said earlier, it is a search that must be
carried out with the naked eye. And if under favorable conditions somebody succeeds in detecting the clouds and wishes to photograph them, he
will not have serious instrumentation problems. The most reliable instrument, in fact, is a short focal length (f = 50 mm), high luminosity (f/1.5
or f/2 aperture) camera. The camera should, of course, be mounted
on a tripod, since a sufficiently long exposure must be obtained.

DUST RINGS

While the question of the satellite clouds in the Lagrangian points was
being debated, and attempts were made to verify it, Kordylewsky com-
municated another discovery to the Belgian astronomer S. Arend, who
subsequently made it public: small earth satellites were not clustered
only in the regions of the two clouds but along the entire length of the
moon's orbit as well.

Discovery had occurred during a series of visual observations carried
out on fifteen nights between October 9 and December 15, 1966, on the
Polish ship Olesnica which was cruising around Africa. Ten members
of the expedition independently observed about a thousand small
nebulosities that, on the basis of careful measurements, appeared to be
distributed all along the moon's orbit (figure 2.2). Thus in addition to
the moon the Earth seemed to be endowed with a whole swarm of minute satellites clustered in a ring similar to that of Saturn.

These findings were received with the same degree of skepticism elic-
ited by the discovery of the clouds, and even now are not generally accepted. Nevertheless, independent of this discovery, and actually prior
to it, other astronomers had found some evidence that the earth may
have an infinite number of microsatellites, most of which would be concentrated in a ring similar to that observed by Kordylewsky, though very
much smaller. However, none of these results have yet been confirmed.

CYRILLIDS AND TEKTITES

Repeated efforts have been made in the past to look for natural earth satellites of more substantial size that describe nonlunar orbits. For a brief moment the search appeared to have met with success.

In 1969 J. P. Bagby claimed to have found about ten satellites that in his opinion had originated from a larger body which had broken up at the end of 1955. In 1973, however, J. Meeus demonstrated that claims for the existence of these phantasmal moonlets, as he called them, were totally unfounded. His arguments against Bagby's alleged proofs were truly overwhelming.

Yet there are sounder though more puzzling observations pointing to the fact that small earth satellites might indeed exist, though perhaps transiently. A supporter of this theory is J. A. O'Keefe who reopened a very unusual case in a scientific paper published in 1959.

On February 9, 1913, at 9:05 P.M., a number of astronomers (including C. A. Chant) observed some strange celestial objects crossing the sky over Toronto. First they spotted a bright red object approaching from the northwest and moving parallel to the horizon. Then, as they were still marveling at this apparition, more objects in small groups of two, three, and four appeared from the same direction and followed the first along the same path. They all had small bright tails that made them look like rockets (figure 2.3). At the end of this apparition the astronomers distinctly heard a booming sound, as of distant thunder. The entire phenomenon did not last more than three and a half minutes. The swarm was observed not only in Toronto but also in several other places along its trajectory on an arc reaching down to Cape St. Rocco in Brazil. From all these observations astronomers computed the orbit of the objects for the period over which they had been observed and found it to be circular. More accurately, according to O'Keefe, the objects described an orbit of zero eccentricity, with an inclination to the terrestrial equator of 50.6° and a period of 90 minutes. Since they had not fallen to earth on a parabolic trajectory, but at least at the time of observation were moving in a circular orbit, they had to be considered satellites of the earth. It would be interesting to know therefore whether the swarm was observed again on its next passage. An hour and a half after appearing over To-

tektites from the same region have the same age, while those from another area may be of a different age. For example, tektites found in Australia and Indochina are only 700,000 years old, while those found in the Moldava basin are 15 million years old, and those from North America originated between 30 and 35 million years ago.

All known facts seem to point to the extraterrestrial origin of tektites. But which regions of space did they come from? And why are they so different from meteorites?

The first clue toward the solution of the problem of origin came from measurements of the aluminum isotope Al 26. This isotope is produced by cosmic rays, to which tektites would have been exposed had they come from space. Furthermore, from the amounts of it can be calculated how long the tektites were exposed to cosmic rays, that is, the length of time they spent in space. Al 26 was indeed found in tektites, but in such small quantities that their space flights could not have lasted more than 10,000 years. This was positive proof that they had in fact come from space, although from regions quite close to earth.

In the meantime laboratory experiments were making another significant contribution to the understanding of tektites. About 15 years ago, in the course of experiments with spheres of glassy material similar to tektites, scientists were able to determine that the latter had been formed by the fusion of bodies entering the earth's atmosphere at grazing incidence and with velocities ranging from 6.5 to 11.2 km/s. These velocities are substantially smaller than those of meteorites (from 12 to 70 km/s). Another significant discovery was that tektites showed signs of successive fusions.

On the basis of all these findings, scientists could now attempt to explain their origin. A number of hypotheses were proposed, including the possibility that they might be remnants of natural earth satellites. Increasing evidence, however, indicated that these bodies were ejected from the moon as a result of meteorite impact. Given the low escape velocity from the moon (2.4 km/s) following the impact of a large meteorite, fragments of the lunar surface can be expelled with sufficiently large velocities to avoid recapture. Some of these may also be injected into trajectories that reach the earth. The ejected material, which is fused at

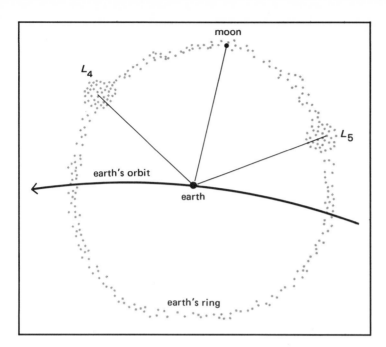

Figure 2.2 Matter distribution along the lunar orbit according to Kordylewski. This material forms a ring around the earth much thinner than that of Saturn and barely visible. Two clouds of condensed dust fall at the Lagrangian points. (From *Ciel et Terre*.)

Figure 2.3 The Cyrillid shower observed in the sky over Toronto during the night of February 9, 1913, as represented in a painting by Gustav Hahn who witnessed the event. (Dunlap Observatory.)

ronto, having already completed an entire revolution around the earth, it should have crossed the skies over Nebraska, Iowa, and Missouri. Although the night of February 9, 1913, was very clear, the strange phenomenon was not observed in any of these three states. O'Keefe was then led to conclude that the swarm of satellites was on a fall trajectory when observed from Toronto to Cape St. Rocco, and that it had hit the ground soon after appearing in the sky. The various objects making up the swarm were called Cyrillids; since nobody had understood what they were, it seemed best to name them after the saint celebrated on the day of their appearance (St. Cyril). Had they appeared today, they would be promptly called UFOs.

Since the 1913 disintegration and fall of a natural earth satellite appeared to be a unique event, O'Keefe's research might have ended right there. A few years later, however, O'Keefe reopened the case by claiming to have found the remnants of other falls; in his opinion they were fragments of satellites that had existed in the past. As it happened, the study of these fragments led to far more unexpected and intriguing results.

In certain regions of our planet strange glassy stones can be found that have proven a considerable puzzle to geologists. They are brown or greenish in color, rounded in shape, and look somewhat like buttons, drops, or small rods. As a rule their diameters are smaller than 2 or 3 cm, although some have been found that are larger than 10 cm. They are rich in silica and aluminum but poor in iron, and altogether might be said to resemble certain volcanic silicates. The strangest thing about them is that they have no geological connection with the terrain in which they are found. In 1900 F. E. Suess gave these strange minerals the name of tektites.

They can be found only in a few well-defined regions of the earth: Indochina, Australia, the Ivory Coast, Texas, Czechoslovakia, namely (figure 2.4). All tektites found in the same area are similar in structure and composition and have the same age. It is possible to determine the number of years elapsed since the time of formation by measuring the amount of argon produced by the disintegration of a radioactive isotope of potassium. With the help of this technique scientists ascertained that

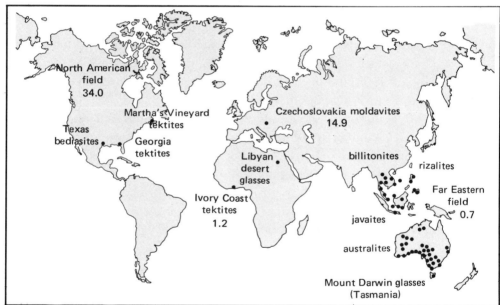

Figure 2.4   *Above*, tektites from the collection of the Specola Vaticana at Castel-gandolfo. (Courtesy of the Specola Vaticana.) *Below*, distribution of tektites on the earth's surface. Indicated are the different names given to them by geologists and the ages in millions of years.

the time of the meteorite impact, cools during its journey to earth and fuses a second time during entry because of attrition with the atmosphere. Accordingly, each group of tektites would correspond to the formation of a large lunar crater by the impact of a specific meteorite. This would explain why tektites from the same area are of the same age and composition, and why they are so different from those of other regions.

This hypothesis was confirmed by the experimental and theoretical work of D. R. Chapman on the Australian tektites, and also by a comparison between the composition of tektites and that of the lunar soil. The Australian tektites have been found in about two hundred different places, all located along a sort of S-shaped belt extending from Australia to Indochina. A variety of these tektites was found to be remarkably similar in composition to the rocks located at the edge of the lunar crater Tycho, which was examined by the U.S. probe "Surveyor 7" in the January 1968 landing. This important discovery spurred scientists to continue investigations in this direction.

After computing thousands of possible trajectories radiating from Tycho, Chapman found that the entire tektite belt from Australia to Indochina could have been formed by a shower of projectiles originating from Tycho, and traveling in the direction of one of the crater's famous rays, that which crosses crater Rosse (figure 2.5). The age of the Australian tektites is quite consistent with selenological evidence of the relatively recent formation of crater Tycho.

Since we know that the Australian tektites are 700,000 years old, it follows that the great crater Tycho was created only 700,000 years ago by the impact of a giant meteorite. We have also learned that should another large meteorite strike the moon and form a huge crater, the earth would once again be hit by a shower of fiery drops.

THE PLANET VULCAN

From the beginning of the eighteenth century Mercury has been somewhat of an enigma. Since its orbit lies inside that of the earth and the planet revolves between us and the sun, at times it will pass right in front

Figure 2.5   The full moon photographed at Mount Wilson with the 254-cm telescope. Clearly visible in the lower part are Crater Tycho and the large number of rays originating from it. Tycho is believed to have formed 700,000 years ago from the impact of a large meteorite. The Australian tektites, which are as old as Tycho, would have been ejected from it along one of the rays (that crossing crater Rosse). (The Hale Observatories.)

of the sun, appearing as a black dot that crosses the solar disk in a few hours. If Mercury's orbit were in the same plane as that of the earth, this event would take place three times a year (every time the planet passes between us and the sun). As it happens, Mercury's orbit has an inclination of 7° to the ecliptic plane and therefore almost every time the planet comes between us and the sun it passes slightly above or below the solar disk. Occasionally, however, the sun, Mercury, and earth are in perfect alignment. This occurs when Mercury's position is close to the node, that is, to one of the two points where its orbit intersects the plane of the ecliptic. At this point the planet is in the same plane as the earth, and we can see it projected on the sun's disk. This phenomenon occurs somewhat irregularly at intervals of 13.7, 10, or 3 years.

The same phenomenon should occur for all the planets that revolve between us and the sun, though less frequently the farther the planets are from the sun. Venus is the only other planet we know that lies between us and the sun, and it does in fact periodically transit the solar disk at intervals of 8, 121.5, 8, 105.5, 8, 121.5, and so on, years.

Due to Mercury's distance and small size, its transits can be observed only by telescope (figures 2.6 and 2.7). The oldest known transit was predicted by Kepler and observed by Pierre Gassendi in Paris on November 9, 1631. Henceforth transits were observed fairly regularly and proved to be quite useful for comparing Mercury's real circumsolar motion to that predicted by the ephemerides, which could then be corrected accordingly. Tables of planetary motions had been successfully computed on the basis of Newton's law of universal gravitation, but Mercury did not appear to conform exactly to the rule. For instance, P. de La Hire's prediction that it would transit the sun on May 5, 1707, turned out to be in error by one day; the May 6, 1753, transit (computed by La Hire and Halley) missed by several hours, and not even J. de Lalande, who corrected the tables for the 1789, 1799, and 1802 transits, was able to produce a complete theory of Mercury's motion.

The problem was studied in depth by U. Le Verrier, who in 1846 had received worldwide acclaim for his discovery of the planet Neptune. He had accomplished this at his desk, without having sighted the planet, inferring its existence from the irregularities in Uranus's motion. After

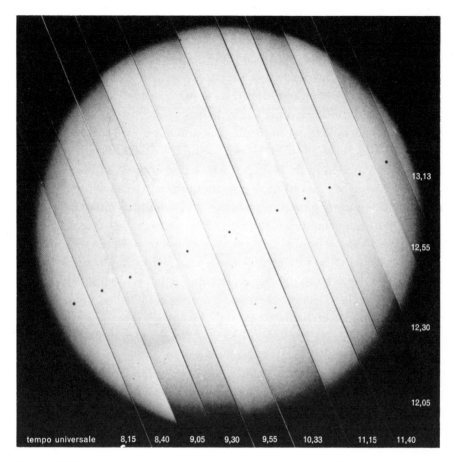

Figure 2.6   Mercury's transit over the solar disk photographed on November 10, 1973, by G. Klaus. Hours and minutes are given in universal time (UT). (Courtesy of G. Klaus.)

studying all the transits observed until then, Le Verrier found that Mercury's perihelion (the point on its orbit closest to the sun) exhibited a more rapid displacement than that predicted by Newton's theory. He computed the residual perihelion advance to be 38 arc seconds per century. In 1849 Le Verrier announced his results to the Academy of Sciences, adding at the same time that he had as yet been unable to conceive of a theory that could account for this phenomenon. In 1859 Le Verrier announced that he had found the explanation: the residual advance of Mercury's perihelion could be explained by perturbations caused by a planet (or group of planets) revolving about the sun in an orbit smaller than that of Mercury. This new planet could be detected either by exploring the environs of the sun during total eclipses, or, as in the case of Mercury and Venus, by observing its passages in front of the solar disk.

Upon hearing about Le Verrier's theory, Dr. Lescarbault, a physician who lived at Orgères and dabbled in astronomy in his modest private observatory, decided to make known an observation he had made on March 26, 1859, which he thought might be of the greatest significance. On December 22 of the same year he wrote to Le Verrier informing him that on that date he had observed a round object of very small diameter (about one-fourth the diameter of Mercury, as he remembered it from the 1845 transit) passing in front of the sun. Wild with joy, Le Verrier rushed to Orgères and addressed Lescarbault with words to this effect: "Sir! Are you the man who claims to have found the intramercurial planet and who has shamefully kept it a secret for nine months? I warn you that I have come to expose your presumption and to discover whether you are deluded or dishonest." Lescarbault explained what he had seen and showed him his instruments, which in truth were rather modest and primitive. Le Verrier demanded to be shown the journal of his observations, and Lescarbault simply replied that it did not exist. Since he was always short of paper, it was his habit to jot everything down on a sort of blackboard and then erase it. For lack of better records, he reconstructed his observation from memory. Le Verrier declared himself satisfied. He estimated that the new planet had a mass 1/17 that of Mercury and a revolution period of 19.7 days. He named it Vulcan.

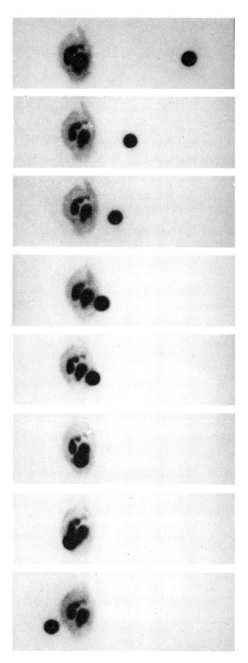

Figure 2.7 The passage of Mercury across a sunspot observed from Catania on May 9, 1970. The photographs were taken by the German amateur astronomer H. Bernhard with the finding telescope of the 90-cm telescope at Catania Observatory. (Courtesy of H. Bernhard.)

That Le Verrier should have accepted so readily such scanty evidence is little short of unbelievable. Perhaps this was an extreme case, but it was neither the first nor the last time that a theoretician, unskilled at observational work, would enthusiastically endorse unreliable observations whose only virtue resided in their being in perfect agreement with his own pet theories.

In the meantime, news of Le Verrier's latest achievement had crossed the channel and had been enthusiastically received in England. It came as a shock to the London city chamberlain, B. Scott. In a letter to the *Times*, appearing January 10, 1860, Scott let it be known that he had already discovered the intra-mercurial planet in midsummer 1847, when he had observed a body the size of Venus transiting the sun. He had sent word of his discovery to a member of the Astronomical Society, a Mr. Abbatt, who had answered that in his opinion it was just a sunspot. Two days later Abbatt confirmed Scott's claim in the *Times*, and expressed his regrets at having unwillingly deprived him of priority in the discovery. However, Scott's planet could not have been the object seen by Lescarbault, which was much smaller. In itself, this was not a bad thing, since the mass that Le Verrier had estimated for Vulcan was too small to account for the perturbations in Mercury's orbit. In effect, there were not one but two possible intra-mercurial planets.

The director of the Zurich Observatory, R. Wolf, also became interested in the new planet, and promptly set out to find some evidence for it in solar observations of the previous century. He made a list of twenty-one possible sightings of the intra-mercurial planet and transmitted it to Le Verrier. After computing the positions of the nodes, it appeared that the new planet would transit the sun between March 29 and April 7, 1870. A worldwide search was instituted during that period, but the planet could not be found.

In the meanwhile, Lescarbault's observation had come under criticism by C. Flammarion, and C. H. F. Peters. But the strongest objections came from E. Liais, a French astronomer working in Brazil, who stated that on March 26, 1859, he had also observed the sun but had seen no trace of the hypothetical planet.

Despite these doubts, Vulcan's supporters were heartened by some new observations. On March 20, 1862, the English amateur astronomer M. W. Lummis observed a body transiting the sun whose revolution period was computed at 17.5 and 19.9 days by two different astronomers. Both Peters in the United States and Spörer in Europe were positive that it was a sunspot they themselves had observed, but Le Verrier and other supporters of Vulcan regarded the new observation as further evidence of their theory. Together with a similar sighting made in Constantinople on May 8, 1865, it began to lend support to the existence of Vulcan, which by then had found its way into a number of textbooks.

In 1872, the superintendent of the *British Nautical Almanac* decided to re-examine the whole question. After reviewing Le Verrier's calculations and the most reliable observations made to date, he concluded that Vulcan should transit the sun on March 24, 1873. Observations made in Europe, America, and Australia revealed nothing at all. These negative results dismayed even the most passionate of Vulcan's supporters. In April 1876, however, interest was rekindled by the news that on the 4th of the month a round object had been seen transiting the sun from the town of Peckaloh. Wolf found that its orbit appeared to agree with that of Vulcan and informed Le Verrier. Assuming a period of 42 days, the April 4 transit would have been the 148th after that observed by Lescarbault. Le Verrier then reopened the investigation of the entire question, drew up a list of all transits observed from 1802 to 1876, excluding the unreliable ones, and divided them into four groups, each corresponding to a possible planet. The most promising group included five observations, all apparently relating to a single object with an estimated period of 33 days. The next possible transit was expected on the 2nd or 3rd of October 1876, but no planet was seen. Le Verrier asked for additional observations on the 9th and 10th, but once again the results were negative. Next it occurred to him that Vulcan's orbit might be inclined to the ecliptic plane by as much as 10.9°. In this case a transit should be observed on March 22, 1877, while the next one would not take place until October 15, 1882.

In the meantime the April 4 transit had been dismissed as a sunspot unaccompanied by penumbra which had been observed in Spain and even photographed at Greenwich. Le Verrier conceded the point. Despite the setback, hope for a definitive discovery spurred astronomers, particularly amateurs, to continue their efforts, and everyone with a telescope searched for Vulcan. Many indeed found it, or claimed to have found it in the past, and many letters to this effect were sent to the *Scientific American* where for a while they were duly published, although at the end of 1876 the journal stopped printing them.

When the fateful date of March 22, 1877, drew near, Le Verrier sounded a new alert and suggested to astronomers that they should keep watch on the sun on the preceding as well as the following day. For three days telescopes were trained on the sun, but again nothing was seen.

That same year, on September 23, the principal character in this drama passed away. Le Verrier, the first man to discover a planet by mathematical deduction alone, without having seen it and apparently without ever seeing it in his life, ended his days still believing that his calculations had led to the discovery of a new planet in the innermost part of the solar system. With the death of Le Verrier enthusiasm abated very rapidly, at least in Europe.

In the United States, on the other hand, faith in Vulcan had not been completely shaken. After the failure of so many attempts to find Vulcan by the method of transits, a number of astronomers decided that they might do better by exploring the sun's environs at times of total eclipse. On July 29, 1878, while the sun was occulted by the dark disk of the moon, several telescopes pointed to the region around it. Some famous and experienced astronomers like E. J. Holden did not see anything, but two others claimed to have discovered Vulcan. They were by no means inexperienced: one of them was J. C. Watson, director of the observatory at the University of Michigan and the discoverer of twenty-three small planets; the other was L. Swift, a skillful amateur who throughout his life discovered twelve comets and hundreds of nebulae. Watson was the first to announce a sighting. On the return of his expedition to observe the eclipse, on August 1, he claimed to have seen Vulcan 2° southwest of the sun, about a minute before the end of totality. The planet appeared of magnitude 4.5. Its position did not correspond to that of any known star, and in any case its image through the telescope was not starlike. Shortly afterward came Swift's announcement which gave the planet's position relative to the star Theta Cancri.

The press reacted with great enthusiasm to the news of this discovery, which at first was not challenged even in scientific circles. On August 1 J. N. Lockyer wired the news to G. Airy at Greenwich. Airy answered on the 3rd suggesting that the object might be the star Theta Cancri itself, whose coordinates were almost identical to those inferred for Vulcan, but both Watson and Swift pointed out that they had used that very star as a reference object. An assistant of Le Verrier, A. Gaillot, found that the observations agreed fairly well with the most acceptable of the orbits proposed by Le Verrier in the past, and A. Mouchez, director of the Paris Observatory, announced that Vulcan's existence appeared now almost certain and that its discovery was the crowning glory of Le Verrier's scientific career.

Toward the end of August, however, the situation became less clear. Watson found that the position obtained by careful analysis of the observations did not coincide with the one he had first announced and therefore was no longer in agreement with Swift's position. Thus for the second time there appeared to be two planets where one was expected. Peters reopened the polemic against Vulcan by sharply criticizing both Watson's and Swift's observations. Even though both observers were quite experienced, it must be remembered that observing conditions during a total solar eclipse are very difficult and quite different from nighttime conditions for two reasons: darkness comes very suddenly, and the time available for observations is quite limited; in that specific case it had been 2 minutes and 40 seconds. Naturally Watson and Swift answered their critics in defense of their own positions, and a polemic ensued that involved Peters and Watson until the latter's death in 1880, shortly after he founded a special observatory for the search of Vulcan at the University of Wisconsin.

Most other astronomers decided to ignore the controversial observation of 1878 and began to hunt anew during the solar eclipses of May 17, 1882, and May 6, 1883. Subsequently, C. D. Perrine assumed observations during the eclipses of 1901, 1905, and 1908 and obtained deep photographs of the sun's environs which could detect stars as faint as magnitude 9. Later on, W. W. Campbell and R. J. Trumpler were able to photograph stars as faint as magnitude 10 and at a distance from the sun up to 10° (Mercury's perihelion distance is 18°), yet found no trace of intra-mercurial planets. In addition, Trumpler carried out a photographic search along Mercury's orbit at the Lagrangian points L_5 and L_4, and he concluded that in those two regions there could be no planets with diameters larger than 20 and 60 km, respectively.

After so many fruitless investigations and contradictory results, it is not surprising that there should follow a period of deep skepticism bordering on total indifference to the whole question. Yet, even if the planet did not exist, the displacement of Mercury's perihelion still remained an undeniable fact; furthermore, better observational techniques had pushed the value of the residual displacement from 38 to 43 arc seconds per century, as shown by S. Newcomb in 1882.

Even if Vulcan did not exist, there still had to be some reason for this displacement. In 1885 Newcomb suggested a number of hypotheses that could account for the phenomenon. They can be summarized as follows:

1. deviation of the sun's shape from a perfect sphere,
2. a ring of small intra-mercurial asteroids, rather than a single planet,
3. a significant amount of diffuse matter,
4. a system of small planets between Venus and Mercury,
5. an imprecision in Newton's formula.

It would serve no purpose here to discuss the weaknesses of each hypothesis, particularly in view of the fact that we have now come to the final act and the conclusion turned out to be entirely different from what anybody had predicted.

At the end of 1915, having completed his general theory of relativity, Albert Einstein found that the theory required an excess in the advance of Mercury's perihelion of 42.91 arc seconds per century; the most exact value obtained by observation was 42.84 arc seconds per century.[2] Thus

2. In 1975 L. V. Morrison and C. G. Ward, after examining 2,400 observations of Mercury's transits from 1677 to 1973, found a slightly lower value, 41.9 arc seconds ± 0.5.

the new theory not only explained but in fact required the excessive value for the displacement of Mercury's perihelion which had been found.

At this point the question of Vulcan should be considered definitively closed, but it is not so. On the one hand, hypotheses have been put forth that would explain the displacement of Mercury's perihelion without invoking general relativity; on the other, even explaining the displacement on the basis of general relativity, there is still the possibility that there might exist one or more intra-mercurial planets of small mass.

As a matter of fact this question has been recently revived as a result of the photographs made by H. C. Courten during the solar eclipses of 1963, 1966, 1968, and 1970, which indicate the existence of still unidentified celestial bodies in the proximity of the sun. Of course, one should add that these bodies merely appeared to be near the sun and that their closeness might have been only a matter of perspective. In other words, they might have been very distant objects, such as small planets or comets, which by pure chance appeared to us at that particular moment to be projected on a region of the sky quite close to the sun.

But the doubt remains that they could be objects relatively close to the sun: and so the phantom of Vulcan returns to haunt us again.

PLANET X

In 1972 to 1973 radio, press, and television the world over announced the discovery of a tenth planet of the solar system traveling in an orbit outside that of Pluto, the farthermost known planet. It was promptly dubbed Planet X, a play on words based on the fact that X means "tenth" in Roman numerals and is also the symbol of mystery in every classic thriller.

The existence of a trans-plutonian planet had been suspected immediately after the discovery of Pluto itself. Before Uranus was discovered, nobody had ever imagined that there might be more planets than those known from antiquity, that is, in order of distance from the sun, Mercury, Venus, earth, Mars, Jupiter, and Saturn. In 1781 W. Herschel accidentally discovered Uranus, and henceforth nobody could any longer be certain of knowing the whole of the solar system. There was reason to suspect that another planet might exist beyond Uranus's orbit, and in fact Neptune was discovered in 1846. A few years later speculation began about the possibility that there might be yet another planet beyond Neptune, and in 1930 Pluto was discovered.

The two outermost planets were discovered by studying the perturbations caused by their masses on Uranus's motion. Speculation on the next planet started from a study of the grouping of cometary orbits. As we have seen, the gravitational pull of the planets can change the orbits of many comets into very extended ellipses. Consequently, different families of comets develop, each bound to a planet. The average maximum distance from the sun (aphelion distance) of each family is slightly larger than the aphelion distance of the parent planet. Except for Saturn's family, this difference is 10 percent of the planet's aphelion distance.

You will recall that in 1950 Schütte assigned 52 comets to Jupiter's family, 6 to Saturn's, 3 to Uranus's, 8 to Neptune's, and 5 to Pluto's. No planet had ever been observed beyond Pluto, but Schütte noted the presence of a family of 8 comets with a mean aphelion distance of 85 AU. From the existence of this family one could infer the presence of a planet that according to the empirical rule of 10 percent should be located at a distance of 77 AU from the sun. This tenth planet would complete a revolution about the sun every 675 years. Since the aphelion distances of the farthest known comets differed very much from one another, no additional comet family could be easily recognized beyond the last one. Thus, according to Schütte, the trans-plutonian planet had to exist, and at the same time had to be the farthermost in the solar system.

More than twenty years went by, and the tenth planet was not discovered. In the spring of 1972, however, J. L. Brady published the type of news that in the hands of newspapermen is apt to become an immediate sensation: not only did he claim that the tenth planet existed, but that he had computed its orbit and mass, as well as a position on the sky where it could be found and observed.

It had all started with a study of Halley's comet. On the basis of five thousand observations obtained during the comet's returns of 1682, 1759, 1835, and 1910, and taking into account the perturbations exerted by all nine major planets of the solar system, Joseph Brady and Edna Carpenter had computed with great precision the circumstances of the comet's next apparition, predicted for 1986. Furthermore, by introducing a corrective term in the equation of motion, they could improve the precision of the fit to the orbital elements of all the comet's passages, starting from that of 86 B.C. According to Brady, the physical interpretation of the corrective term derived from the fit was the existence of a trans-plutonian planet, or Planet X.

Up to this point Brady's suggestion was not very surprising. This hypothesis had been put forth by both Le Verrier in the case of Neptune and P. Lowell in the case of Pluto and had in fact led to the discovery of the two planets. What struck astronomers as being very strange in Brady's announcement were the characteristics he had assigned to the new planet: an aphelion distance of 59.9 AU and an orbital period of 464 years. Quite extraordinary also was the inclination of its orbital plane, 120°. This meant that the planet was moving in the opposite direction from that of all the other planets in the solar system and that its orbit had an inclination of 60° to the plane of the ecliptic, where practically all the other planets lie. Even more startling was its mass, estimated to be three times larger than that of Saturn, one of the most massive planets in the solar system. Such a massive planet had to be very large and therefore should have appeared brighter than Pluto. The fact that it had never been seen, was very strange indeed.

In any case, since Brady had furnished the position on the sky where the planet was to be found, this difficulty could soon be remedied. The first photographic search was carried out at the observatory of Herstmonceux in the summer of 1972. The result was negative. In the region of its estimated position there was no planet brighter than magnitude 15.5, while Planet X should have been at least brighter than magnitude 2. An independent search was conducted at the Lick Observatory. Planet X could not be found even though it should have been a hundred times brighter than the faintest stars photographed in this survey. As if these

negative results were not enough, a few months later several astronomers proved that the planet could not exist.

A first argument along these lines was developed by P. Goldreich and
W. R. Ward, who demonstrated that a planet with the characteristics
suggested by Brady would have disrupted the structure of the solar system from Jupiter outward through its perturbations. Since this process
would require about a million years, we would see its effects by now; obviously there is no reason to think that Planet X has only recently come
into being. Additional studies carried out by P. K. Seidelmann, B. G.
Marsden, and H. L. Giclas provided definitive proof that Planet X did
not exist. In fact, should there be a planet with the mass and orbit suggested by Brady, the motions of Jupiter, Saturn, Uranus, Neptune, and
Pluto would appear quite different from those we observe. In particular,
they should show yearly displacements of their positions as large as 30
seconds, and such large values would have been noticed long ago.

One final proof was added to the various arguments against the existence of Planet X. In 1971 T. Kiang reconstructed the past orbits of
Halley's comet on the basis of ancient observations, for the most part
Chinese, some of which had never been published. He was able to determine the dates of the comet's perihelion passages in the past centuries
with greater precision than had been done previously.

Not only were they not consistent with the dates computed by Brady
and Carpenter, but the difference increased the farther back in time
one went.

Consequently, the orbital variations of Halley's comet could not be explained by perturbations caused by a tenth planet, as was Brady's con-
tention. It was undeniable, however, that the actual times of perihelion
passage were different from the estimated dates, and this discrepancy
still remained to be explained. In Kiang's view, an effect on the comet's
motion exactly equivalent to that of Brady's planet could be caused by a
series of impulses tangential to the motion of the nucleus at the time of
perihelion passage. As we have seen, this acceleration has been found to
occur in a number of comets, and in Whipple's model it is explained as a
jet reaction due to the very structure of the comet's nucleus and to the

violent action of the sun's heat on it. In sum, the reasons for the orbital variations of Halley's comet are not to be found in a distant planet that nobody has been able to see but in the comet itself. While it certainly obeys the law of universal gravitation, it is also true that at each perihelion passage the comet alters its own orbit under the influence of non-gravitational forces that are triggered by the sun and develop from its interior as from a spaceship.

The question of the tenth planet can of course be considered from a different point of view. Even though Brady's planet does not exist, there could still be another planet beyond Pluto, as suggested by the comet family found by Schütte. This problem was re-examined in 1973 by D. Rawlins and M. Hammerton through a study of Neptune's residuals. Their conclusion was that an additional planet might exist, although of small mass and low brightness not exceeding magnitude 17. Thus the question of the trans-plutonian planet is still open.

3   THE KILLER MONSTERS

THE GUM NEBULA

If our eyes were as sensitive as the optical instruments used by astronomers, in addition to stars we would be able to see a great variety of objects in the night sky, and among them hundreds of gaseous spheres of different colors and sizes scattered here and there like so many toy balloons. One of these spheres is enormously larger than all the others, and its edges extend to the vicinity of the solar system itself. Until about twenty years ago nobody—not even astronomers—knew of its existence. It is so tenuous that it can not be detected on normal photographic plates, and so extended that it is necessary to piece together many photographs taken with large-field instruments to obtain a panoramic view capable of revealing its full extent.

If it were luminous enough for our eyes to see it, so that we could appreciate its size, structure, and color, it would appear as a truly awesome sight: a vast, tenuous, reddish cloud, nearly elliptical in shape, partly transparent, and partly formed by veils and wispy filaments, looming in the southern skies over a larger area than that occupied by Ursa Major and Ursa Minor together. Observations have revealed that this ruddy apparition is a giant hydrogen cloud 2,400 light-years across, the largest known in the entire Galaxy. Even though very tenuous, it contains enough matter to form 180,000 stars as massive as the sun.

This nebula appears so large to us not only because of its intrinsic size but also because of its proximity to earth. It is so close, in fact, that one might expect it to engulf the solar system and mankind as well in a not too distant future. This could easily happen if the nebula were expanding or if its bulk motion were directed toward the earth. According to the latest views, however, this possibility can be excluded because not only is the nebula not expanding at present, but perhaps it never did. It is only a fleeting apparition, a ghost from the past that has come to tell us the story of a cataclysmic event that occurred almost on our doorsteps more than ten thousand years ago.

GASEOUS NEBULAE

The solution of the mystery of the Gum nebula is tied to two very different types of celestial objects—gaseous nebulae and supernovae.

Gaseous nebulae are very common celestial objects, whose luminosity can be satisfactorily explained on the basis of the theory of atomic radiation emission. We shall describe them briefly, starting from the simplest case, namely, that of planetary nebulae—those multicolored balloons we mentioned at the beginning of this chapter.

The term "planetary nebulae" used in describing these celestial objects is very misleading, since in fact they have absolutely nothing in common with planets. When viewed through the telescope, many of them appear as fairly regular disks of low surface brightness (figure 3.1) and therefore look somewhat like the planets. Through detailed studies carried out primarily in this century, we have learned that a planetary nebula is a sphere of gas that is as large on the average as the entire solar system yet so tenuous that its total mass does not exceed one-tenth that of the sun. At the center of each sphere there is always a star from which the gaseous shell is known to have originated. This is demonstrated by the symmetry of the shell itself, and above all by the discovery that it expands with a velocity ranging from 10 to 30 km/s, depending on the specific case.

These diffuse nebulae do not always appear as filled-in disks but more often as rings (figure 3.2). This can be easily explained as an effect of geometrical perspective, since in the direction of the central star we look through a smaller thickness of the shell than in the direction of the limb. By an analogous effect, the earth's atmosphere appears thicker and more opaque at the horizon than at the zenith. On colored plates these disks, or rings, appear of different hues depending on the distance from the center. The gas is not self-luminous, nor does it glow by simply reflecting the light of the central star but rather through a process of excitation caused by it. The star within the nebula has an extremely high temperature—from 40 to 100,000°C—and therefore emits a large amount of energy, particularly as ultraviolet radiation. The radiation excites and ionizes the surrounding gas, which then re-emits the absorbed

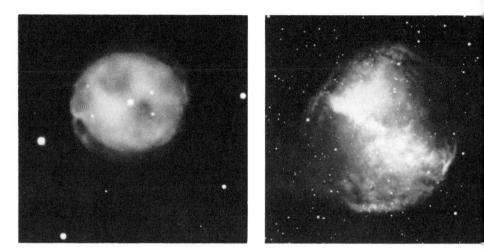

Figure 3.1 Two typical planetary nebulae: *left*, M 87 in Ursa Major; *right*, M 27 in Vulpecula. Photographs obtained at Mount Wilson with the 152-cm and 254-cm telescopes. (The Hale Observatories.)

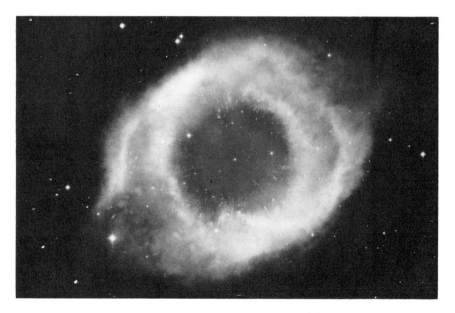

Figure 3.2 NGC 7293 in Aquarius photographed in red light with the 5-m telescope at Mount Palomar. Faintly visible at the center of the nebula is the violet-blue star that excites it. (The Hale Observatories.)

energy as radiation of longer wavelength, corresponding mostly to visible light. Let us explore this matter a little further.

To understand the mechanism responsible for the emission of light, or radiation in general, one must have some knowledge of atomic physics. We will try to explain the bare essentials in the simplest possible way.

According to the model conceived by N. Bohr at the beginning of this century, the atom can be viewed as a miniature solar system, where the center is occupied by the nucleus with positive electric charge, and the periphery by one or more electrons (up to 92, in the case of uranium) whose total negative charge neutralizes that of the nucleus (figure 3.3). Under these conditions the atom is electrically neutral, but if in some manner one or more electrons are removed, it becomes positively charged, owing to an excess positive charge in the nucleus no longer balanced by the missing electrons. In this case the atom is said to be ionized, singly, doubly, and so on, depending on the number of electrons that have been removed.[1]

The electrons revolve about the nucleus according to very specific rules and are restricted to specific orbits that define different energy levels. As long as an electron moves in its orbit, it will neither absorb nor produce energy. But if we wish to transfer it to an outer or an inner orbit, we must supply it with some energy in the first case, or allow it to expend it in the second. Think of it in terms of a ball rolling on the steps of a Greek amphitheater: as long as the ball rolls along the same step it neither absorbs nor expends any energy, and except for attrition, it would keep on rolling forever. If we now wish to transfer it to one of the upper steps, we must supply it with the energy necessary to overcome the change in levels. Should the ball instead start falling down to the lower steps, it would acquire a certain amount of kinetic energy that could be experienced, for instance, by the spectator on the lowest step as he gets hit on the foot.

1. Note that in common terminology the roman numeral I indicates the neutral element, and the subsequent numerals the various degrees of ionization. Thus, for example, O I, O II, and O III indicate, respectively: neutral oxygen, singly ionized oxygen, and doubly ionized oxygen.

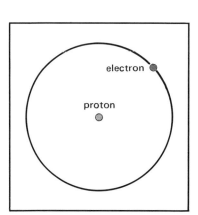

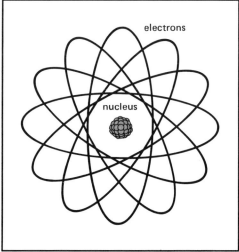

Figure 3.3   Comparison of the hydrogen and uranium atoms. The nucleus is at the center. Hydrogen has only 1 electron, while uranium has 92, a number equal to that of its protons, or atomic number.

Let us consider the case of the hydrogen atom which is the simplest and most common atom in cosmic space, consisting of a nucleus and one electron. To begin with, let us assume that the electron is at the ground level, that is, in the orbit closest to the nucleus (figure 3.4). If we wish to transfer it to the next orbit, we must supply it with a certain amount of energy which is measured in electron-volts (eV),[2] as shown on the right-hand side of figure 3.4. In a more general way, if we want to shift the electron from one possible orbit to another, we must supply it with energy equal to the difference between the energies pertaining to the two levels. The required energy can be provided by light, or radiation in general, and is expressed by the relation $E = h\nu$, where h is Planck's constant, E the energy, and ν the frequency of the radiation. The energy necessary to bring the electron to a given level is called excitation potential. As mentioned above, the energy can be supplied by the absorption of a photon of frequency ν. Conversely, when the electron returns to the original level the difference in energy is radiated as a photon of the same frequency ν.

Once the electron is on a higher level, it does not necessarily follow that after a certain interval it will return directly to the original one. On the contrary, it will generally pass through one or more intermediate levels, emitting at each transition radiation of lower energy than the total jump, and therefore of lower frequency and different color (shifted toward the red). Similarly, electrons may fall to the same level, though coming from different higher levels. Here again, every jump will correspond to a line of different frequency and therefore color. The set of lines obtained in the fall of an electron from different levels to a given level constitutes a series. Thus we have the Lyman series, produced by the transitions from various levels to ground level; the Balmer series, where the final level is 2; the Paschen series, that consists of the transitions to level 3, and so on for Brackett's, where the final level is 4, and Pfund's, where it is 5. The Balmer series is the best known because the

2. The electronvolt (eV) is a unit of energy used in atomic physics. It is defined as the energy acquired by an electron in passing from a point of low potential to a point one volt higher in potential. It corresponds to 1.6×10^{-12} erg.

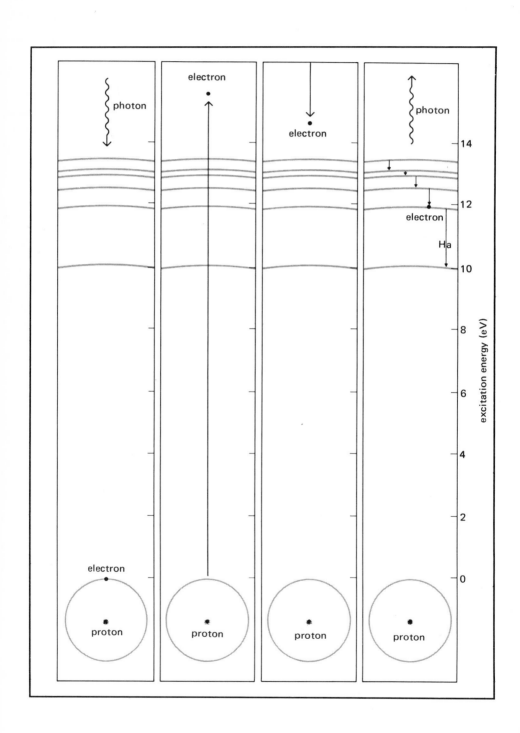

Figure 3.4   Photoionization and recombination of the hydrogen atom according to the classic model. The two columns at the left illustrate photoionization. In the first the neutral hydrogen atom is at the lowest energy level, since the only electron is at level 1 (the closest to the nucleus which is a single proton, also called the ground level). An incoming photon, or radiation quantum, is shown at the top. If the energy of the photon is ≥ 13.6 eV, the hydrogen atom becomes ionized by absorbing the radiation and ejecting the electron (second column). The electron will have a kinetic energy equal to the difference between the photon's energy and the ionization potential of hydrogen (13.6 eV). In the third column the hydrogen ion captures a free electron which has a certain kinetic energy. Starting from a very large distance (infinity) and falling, for instance, to energy level 6 (fourth column), the electron will cause the emission of a photon of a certain energy. This energy is given by the difference between those of the two levels and can be read from the scale on the right. From level 6, the electron can fall to 5, then 4, 3, and finally 2, emitting each time a photon having an energy (and thus a frequency) determined by the difference between the excitation energies of the different levels. In the last jump the electron emits the characteristic H_α red line. Of course, the electron could have fallen to level 2 from 4 (jumping over level 3). In this case it would have emitted a line of higher energy, that is, of higher frequency, or shifted to the violet (H_β). On the other hand, the electron could have arrived at level 2 directly from 5, jumping both 3 and 4, and emitting a third line (H_γ), and so forth. All these lines, formed from a direct jump to a given level from higher levels, constitute a series. In particular, all electron jumps terminating at level 2 produce the Balmer series of lines, H_α, H_β, H_γ, etc. The transitions to ground level produce the higher energy, ultraviolet lines of the Lyman series, L_α, L_β, L_γ, etc. Transitions ending at level 3 give the infrared Paschen series, P_α, P_B, P_γ, etc. The inverse process, which produces absorption lines, occurs when a photon does not have enough energy to ionize the atom but enough to bring the electron from a lower to a higher level. This process is similar to that shown in the first two columns, but with jumps of the type shown in the fourth, although in the opposite direction. (From *Scientific American*.)

energy differences between the various levels are such that they produce lines whose frequencies fall within the visible range of the spectrum. The line produced by the transition from level 3 to 2 is red and is indicated as H_α; the line from level 1 to 2 is blue-green and is labeled H_β; the jump from 5 to 2 produces a violet line, H_γ, and so forth.

This is the mechanism by which an atom emits or absorbs electromagnetic radiation in general and light in particular, that part of electromagnetic radiation with wavelengths between 3,800 and 7,800 Å to which the human eye is sensitive. (Note that 1 Å = 10^{-8} cm = 10^{-10} m.)

Let us assume now that an electron at ground level is supplied with an energy of 13.6 eV or higher. This energy is sufficient to tear that electron completely away from the hydrogen atom, which is then ionized. Similarly, when a hydrogen cloud is hit by the highly energetic radiation emitted by a hot star at its center (as in the case of planetary nebulae), the hydrogen atoms are ionized; hence the gas cloud will consist of ions and free electrons. Occasionally, some of these electrons recombine with ions, and falling to level 2, either directly or after passing through intermediate levels, they cause the gas to emit lines in the Balmer series, in particular the red H_α line, corresponding to the jump from level 3 to 2.

Atoms of other elements behave in much the same way in both the ionization and excitation processes. The only difference is that these processes become far more complex, owing to the greater number of electrons and the many different values of excitation and ionization potentials.

We now have at hand all the elements needed to understand a planetary nebula in all its various aspects, including the variety of colors at different distances from the center. As we have seen earlier, it consists of a more or less spherical cloud of gas, mostly hydrogen, at whose center is a very hot star. This star emits very energetic and intense radiation, whose flux decreases as the distance from the center increases, and as it becomes dispersed into a larger volume. We will consider here only hydrogen, helium, and oxygen, although the gas cloud surrounding the nucleus normally contains other elements also (sulfur, nitrogen, silicon, argon, neon, and so on). These elements are generally found in smaller amounts, and in any case can be discussed in the same way as the first three.

The ionization potential of hydrogen, the energy necessary to remove the single electron from the hydrogen atom, is very low (13.6 eV); so hydrogen can be ionized at great distances from the nucleus. Within this region, the hydrogen atoms cannot recombine, and therefore its protons and free electrons will be incapable of emitting radiation. Energy levels of 13.6 and 35.1 eV are sufficient to ionize oxygen once and twice, respectively, but the ionization potential required to remove a third electron is 54.9 eV. Hence, within the nearly identical shells containing neutral oxygen and hydrogen, there will be a region of singly ionized oxygen, and inside this region will be another consisting of doubly ionized oxygen. This last one will contain also singly ionized helium. Neutral helium has an ionization potential of 24.6 eV, and therefore is more difficult to ionize than hydrogen or oxygen but easier than ionized oxygen. The ionization potential necessary to ionize helium a second time is 54.4 eV, nearly the same as that required to ionize oxygen three times. Thus, as we move away from the central star, the ionizing power of its radiation continually decreasing, we will meet in succession several concentric shells containing O III and He II, O II, He I, and lastly O I and H I.

All this material can emit radiation through three different mechanisms:

1. photo-ionization and subsequent recombination, which is the mechanism that we have already discussed for the hydrogen atom;
2. excitation by electron scattering of the so-called forbidden lines, which are difficult to obtain in the laboratory but are extremely common in the very low-density conditions in which gas exists in planetary nebulae;
3. fluorescence due to random coincidences.

As we mentioned earlier, hydrogen can be ionized at great distances, but there will be a limit beyond which the radiation from the central star is insufficient to ionize it and where the hydrogen will be in the neutral state. Under these conditions it is unobservable, at least optically. But at about that limiting distance there forms a spherical shell where an equilibrium is established between the number of hydrogen atoms that are being photo-ionized and the number of free protons and electrons which recombine to form neutral hydrogen. In this process of recombination the hydrogen atoms emit radiation corresponding to the various lines in the different series, particularly in the Balmer series and in the H_α line, thereby giving the outer regions of the nebula its characteristic red color.

In the regions closest to the nucleus, however, there is a prevalence of O III and of large numbers of free electrons torn from hydrogen atoms through photo-ionization and moving at high speed. These electrons excite especially the forbidden lines of O III, which fall in the green range of the spectrum and thus color that part of the nebula a beautiful green.

It can be readily understood that other lines, corresponding to different colors, are produced by these and other elements, both neutral and ionized, located in spherical shells of different thicknesses and at different distances from the nucleus. All this will produce that ensemble of shapes and colors which constitutes our full view of a planetary nebula.

The various features of a planetary nebula can be better appreciated in color plates taken with different photographic emulsions and light filters (figure 3.5). By using different combinations of emulsions and filters, one can isolate specific lines of the spectrum corresponding to different elements, or to the same element in different degrees of ionization and excitation, and thus determine the distribution inside the nebula of the atoms emitting at a specific wavelength. One might compare these plates to a set of X-ray pictures that show us the structure of the nebula at different depths, and it is only by taking into account all of them that we can view a planetary nebula in its entirety.

What we have just described is the most complete and characteristic case, although not all planetary nebulae are necessarily the same. There are some that do not exhibit the phantasmagoria of colors displayed by M 57 or the Dumbbell nebula but rather a simple and uniform red. This happens for the most part when the center star does not have a very high temperature, and light emission is limited almost exclusively to the low-excitation H_α line.

The mechanism of light emission from the gas in planetary nebulae

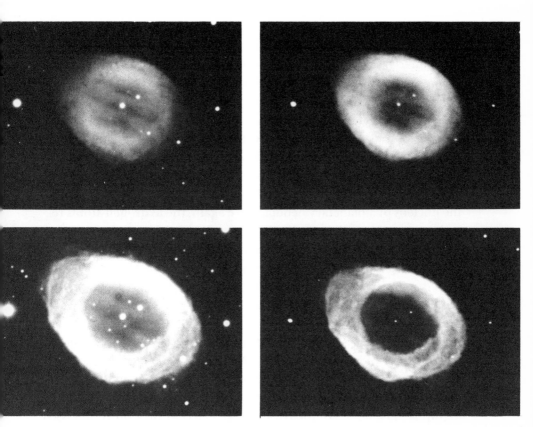

Figure 3.5   The ring nebula M 57 in Lyra in four different colors (from left to right, top to bottom: blue, green, yellow, and red). These photographs were obtained with the 5-m telescope at Mount Palomar. Note that the brightness of the central star decreases from blue to red. The size of the nebula varies in the different colors, depending on the distribution of the elements around it (neutral or ionized) which emit the lines corresponding to the colors of the different photographs. (The Hale Observatories.)

applies also to most of the large nebular systems, such as those in Orion, Trifid, M 8, and many others, where the various constituent elements are similarly excited by one or more hot stars in the central region.

There is a last case which is particularly interesting and of much greater scope. Various kinds of observations, particularly of radio emissions, have revealed the existence of vast interstellar hydrogen clouds, located for the most part near the equatorial plane of the Galaxy. This hydrogen is normally in the neutral state, but, if there happens to be a star, or group of stars, of sufficiently high temperature inside the cloud, all the hydrogen in its vicinity will be ionized as in the regions closest to the nucleus of a planetary nebula. Here again, the hydrogen atoms remain neutral beyond a certain distance. Around this boundary there forms a region consisting of ionized hydrogen, as well as hydrogen in the process of recombining, radiating primarily in the H_α line. B. Strömgren discovered in the 1930s that the ionization process ceases almost abruptly at a certain critical distance from the exciting star. This distance depends on the density of the hydrogen atoms in the cloud and on the spectral type of the star itself (which indicates its temperature). This invisible sphere of ionized hydrogen outlined by a thin red shell is called an "H II region," or "Strömgren sphere."[3] A great number of Strömgren spheres have been discovered to date by means of wide-field photography in H_α light. For the reasons of perspective previously discussed, they appear most often in the shape of a ring, corresponding to the region of maximum optical depth (figure 3.6).

Strömgren spheres immediately suggest an interpretation for the giant nebula we described at the beginning of this chapter. This enormous gas cloud was discovered between 1952 and 1956 by Colin Gum, a young Australian astronomer. Having noticed various nebular frag-

3. The equation relating the three quantities is

$$r_0 = R_0 \, n_H^{-2/3},$$

where R_0 is a constant which can be computed from the star's spectral type; n_H is the number of hydrogen atoms per cm^3 and r_0 the radius of Strömgren's sphere. It is interesting to note that, knowing the spectral type and distance of the radiating star, one can obtain the real value of r_0 and thus derive from the equation the density of the hydrogen cloud.

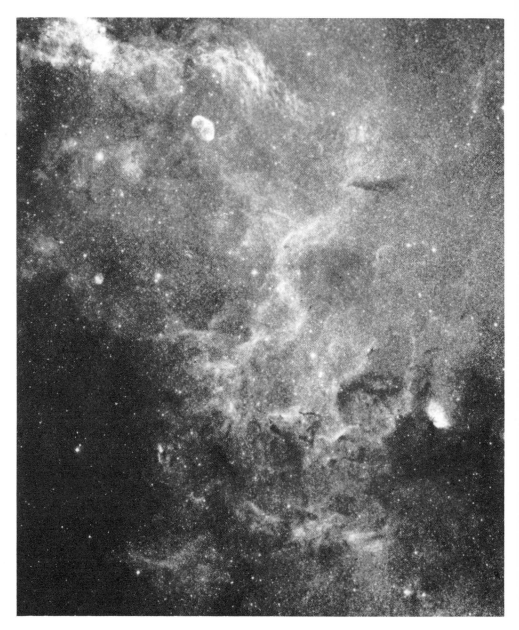

Figure 3.6 The Milky Way in Cygnus with an intricate network of dark and bright nebulosities. A very extended hydrogen emission ring is visible in the upper left-hand quadrant. Inside the ring is an ellipsoidal nebula (NGC 6888) excited by a Wolf-Rayet star. Cyg X-1 is in the lower right of the photograph. (Palomar Sky Survey.)

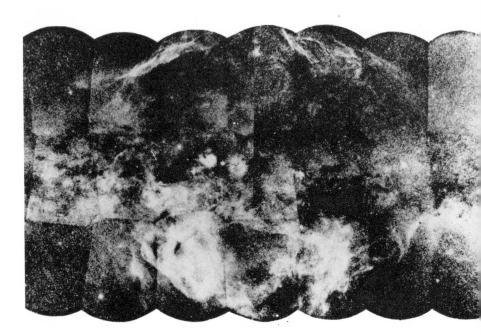

Figure 3.7   The Gum nebula in a mosaic of photographs obtained by the Australian astronomers A. W. Rodgers, C. T. Campbell, J. B. Whiteoak, H. H. Bailey, and V. O. Hunt. (Courtesy of S. P. Maran.)

ments on photographs of adjacent fields, he patiently collected a large number of plates in H_α light and put them all together in a mosaic that gave us the first panoramic view of the great nebula (figure 3.7). Henceforth the object was known as the Gum nebula. The discoverer himself immediately pointed out an interesting coincidence: two hot stars, Zeta Puppis and Gamma² Velorum, were embedded in the central region of the nebula. Gum suggested that their intense ultraviolet radiation could excite the whole nebula, or, to put it in another way, that the nebula might be an enormous Strömgren sphere excited by the two central stars.

Gum's investigations were tragically brought to an end in 1960 by a fatal skiing accident in the Alps. They were resumed in 1970 by a group led by T. P. Stecher and R. Hanbury-Brown, whose work, however, disproved Gum's hypothesis. They found that the temperature of both Zeta Puppis and Gamma² Velorum was around 35,000°K, and therefore too low to excite the whole nebula. It would require an enormous amount of energy (5×10^{51} erg, or as much energy as a star emits in its entire lifetime) to ionize all the gas and enable it to radiate in the H_α line. At that temperature the two stars could have excited only the small region of the nebula closest to them (see figure 3.10).

In the meantime, other astronomers had accumulated very interesting observations pointing to a connection between the Gum nebula and a different class of celestial objects to which we will now turn our attention.

NOVAE AND SUPERNOVAE

At a point in the sky where no star has been previously observed, there suddenly appears a very brilliant star, oftentimes as bright as the brightest in the sky. The star remains at maximum for a few short days, then it begins to fade and in the course of a few months disappears without a trace. The ancients, who could only observe the sky with the unaided eye, were struck solely by the transient aspect of the phenomenon and called these stars "novae" (new), or even temporary stars, because they lasted only a limited amount of time, whereas all the others appeared to be never-changing. However, we now know that if we searched old plates of the region where a nova has appeared, we would find that the star was already there, though so faint as to be invisible to the naked eye. If we continue to photograph the region after the star has become invisible to our eyes, we can observe it getting progressively fainter until it reaches its initial luminosity; henceforth its brightness no longer diminishes. The complete study of the entire phenomenon, including spectroscopic observations, enables us to explain the process as follows.

A star, originally about as bright as the sun, begins abruptly to increase in luminosity until it becomes over a hundred thousand times hotter and brighter. Should there be a planet like ours in the vicinity of the star, as far from it, say, as earth is from the sun, its inhabitants, unaware perhaps of their impending doom, would be blinded and killed in a matter of hours. The star in the meanwhile increases enormously in volume and ejects its outer layers at the astonishing speed of a thousand kilometers per second. In less than two days it would engulf the planet, already shorn of every living thing, and obliterate it completely.

As it keeps on expanding, the envelope cools down, dissipates, and becomes increasingly less luminous and more transparent, until a distant observer, like us, is again able to see through it and view the part of the star immediately below the ejected layer, which is unchanged and has now become its outer surface. The explosion is over. In a few days the star has expended as much energy as the sun emits in 10,000 years and has produced a shell similar to that of a planetary nebula, though smaller and much less dense. According to the most recent estimates, the ejected mass appears to be only one-thousandth that of the sun, a great deal smaller, that is, than the mass of the envelope in planetary nebulae. This does not prevent the cloud from becoming easily visible, since its gases can be excited and radiate by mechanisms similar to those we have described for planetaries. The most spectacular case is that of Nova Persei, which appeared in 1901 and whose envelope is now quite conspicuous.

Although novae are an extraordinary phenomenon, there are still more impressive events, even by cosmic standards. A few novae had been known for a long time as being quite exceptional, such as the 1504 nova observed by the Chinese and the 1572 nova discovered by Tycho

Brahe, both of which became so bright as to be easily visible even in full daylight.[4]

In neither case could the exploding star be detected; but in the first something even more extraordinary was discovered: a rapidly expanding nebula centered on that point in the sky where the star had ignited. Once the expansion velocity had been computed from various observations, the beginning of the expansion was traced back precisely to the year 1054. It was fairly evident therefore that the nebula was the remnant of an explosion that had thoroughly destroyed a star, reducing it to dust in a few hours and scattering its fragments into space.

The study of this nebula revealed some unexpected oddities. In addition to an intricate net of red filaments due to hydrogen emission in the H_α line (which gave the object the name of Crab nebula), astronomers observed a more uniform background, greenish-white, emitting polarized light (figure 3.8). This finding suggested that the nebula might have a very intense magnetic field, along whose lines of force electrons moved in spiral orbits at nearly the speed of light and emitted synchrotron radiation. The Crab nebula was also found to be a source of radio and X-ray emission.[5]

In 1967 pulsars were discovered.[6] They are celestial objects that emit radio pulses of constant amplitude at very short and regular intervals of time. The following year the Crab nebula was found to contain a pulsar emitting thirty radio pulses per second. Soon after it was discovered that the pulsar coincided with a faint star at the center of the nebula whose light pulses with the same period, emitting a flash a hundredfold brighter than the sun every thirty-three-thousandths of a second. Not long after that, it was also discovered that the source emits very short and intense pulses in X rays. After collecting all the observations and develop-

4. As will become clear later in the text, these events correspond to supernova explosions, which involve a completely different physical process.—Trans.
5. S. Bowyer, E. Byram, T. Chubb, and H. Friedman, 1964.—Trans.
6. A. Hewish, S. J. Bell, J. D. H. Pilkington, P. F. Scott and R. A. Collins.—Trans.

ing a coherent theoretical interpretation, astrophysicists concluded that the star had not been totally pulverized in the explosion, that something had remained even in this case. This something had to be a very unusual object—a veritable monster from both the astronomical and the physical point of view—an extremely small object only 30 km in diameter but containing a mass comparable to that of the sun. Considering that the mass of the sun is distributed within a sphere 1,400,000 km in diameter, it is immediately evident that the matter in such an object must be extremely compressed. The density value is so high—a billion tons per cubic centimeter—that we can hardly conceive of it. Under these conditions (or, more accurately, from the time that the density exceeds a hundred tons per cubic centimeter), the peripheral electrons of the atoms penetrate inside the atomic nuclei, where by neutralizing the positive charge of the protons they form neutrons. This is why these objects are commonly called "neutron stars." Because of its small size, a neutron star cannot be observed directly, and its existence can only be inferred from the various effects we have described. Even the light pulses that we observe do not appear to come directly from the surface of the star but from a region about a thousand kilometers away. This region is in rapid rotation around the central body, so that the light and radiation continuously emitted only in one direction strike the earth in periodic flashes at each turn, somewhat like the beam from a lighthouse.[7]

Studies of the pulsars have revealed that the length of the period increases with time. This discovery is very significant in that it provides us with a way to determine how long a pulsar has been in existence, in other words, how long ago the supernova explosion that created it took place. The longest known periods do not exceed 3 seconds and correspond to an age on the order of one billion years. As could be expected, the shortest period is that of the pulsar in the Crab nebula, which is the most recent supernova whose corresponding pulsar has been discovered.

Our examination of supernovae will continue shortly, and we shall then learn that not all of them are quite so typical, but that on the contrary they show individual peculiarities. Right now, at last, we have at

7. For further details on the Crab nebula, see *Beyond the Moon*, pp. 154–159.

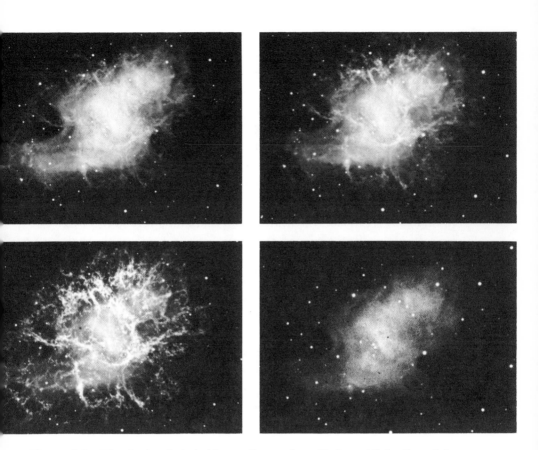

Figure 3.8   The Crab nebula in blue, yellow, red, and infrared light (from left to right, top to bottom). Notice that different regions of the nebula are visible in each color. (The Hale Observatories.)

hand the new elements necessary to solve the mystery of the Gum nebula.

First of all, we should mention that optical and radio observations carried out between 1963 and 1966 revealed the presence of thin filaments in the central region of the Gum nebula which could be interpreted as supernova remnants. In 1968 a pulsar with a very short period (0.09 seconds, or only slightly longer than that of the Crab nebula pulsar) was discovered at their center (figure 3.9). This pulsar therefore must have formed in a relatively recent epoch, though prior to the supernova explosion in the Crab nebula. An X-ray source was discovered in the same region in 1971, but later on it turned out that there were in fact two sources: one small and intense, the other weak and extended. All these findings pointed to the fact that in a not too distant past a supernova explosion had occurred in the central region of the Gum nebula.

In 1971 a group of astrophysicists proposed a new interpretation for the Gum nebula based on the hypothesis that not all the nebulosities visible in this region have the same origin. The net of filaments surrounding the pulsar must have been formed by matter dispersed in the supernova explosion, but the rest of the Gum nebula would be a collection of preexisting nebulosities that could not be optically detected before the outburst because they consisted essentially of neutral hydrogen. Following the explosion, the enormous amount of ultraviolet radiation emitted by the supernova ionized the hydrogen clouds to great distances, and these clouds are radiating now because their protons and electrons are still in the process of recombining (figure 3.10). The most extraordinary thing about all this is that the Gum nebula can be detected today through the light it emits as a result of having been ionized long ago by a source that no longer exists. In a way it acts like a Strömgren sphere. However, it differs from all others known to date in that the central source of ionization has ceased to exist, and for this reason the entire nebula has been called a "fossil Strömgren sphere."

From the length of the pulsar's period it is possible to derive the time of the supernova outburst that created it. It appears that the event occurred about 11,000 years ago, and, in any case, no more than 30,000 years ago. Its distance has been measured by various means and found

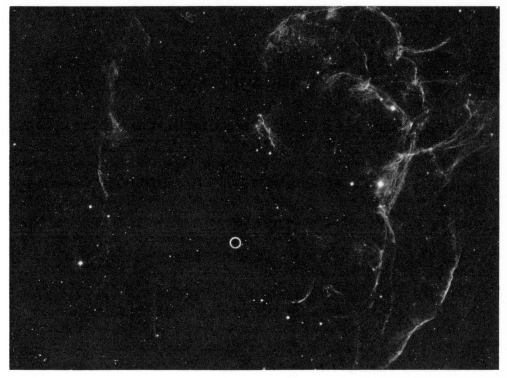

Figure 3.9 A section of the Gum nebula from a photograph made with the Schmidt telescope at Cerro Tololo, Chile. The bright filaments are the remnants of a supernova that exploded at the center of the nebula. The white ring marks the position of the pulsar also shown in figure 3.10 as PSR 0833-45. (Courtesy of B. J. Bok.)

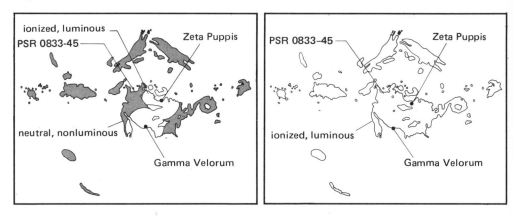

Figure 3.10 Two illustrations showing how the hydrogen in the Gum nebula would appear if it were currently being excited only by the stars Zeta Puppis and Gamma² Velorum (*left*), or if it had been ionized at the time of the supernova explosion (*right*). The luminous regions are distinguished from the neutral and invisible regions which are shown in gray. (From *Physics Today*.)

to be 1,500 light-years, nearly identical, that is, to that of the stars Zeta Puppis and Gamma^2 Velorum which Gum had believed to be responsible for the excitation of the whole nebula. This suggests the possibility that all three stars might have belonged to the same group, although one of them, more massive than the others, evolved faster and exploded.

In view of its distance, the supernova must have appeared in our sky as a star of magnitude -10, or as bright as the quarter moon. It must have been far more impressive, however, partly because of its sudden apparition and partly because all of its light was concentrated in a point source, which made it easily visible in daylight and absolutely dazzling at night. It was certainly seen by prehistoric men who must have beheld it in awe, perhaps in terror. In their ignorance they could only notice the singularity of the phenomenon, but their fears would not have abated had they known its actual cause. Even though they could not know it, they were witnessing at that moment the self-destruction of an enormous celestial object, perhaps far larger than the sun, which released into space almost all the energy contained in its mass.

Today it is all over. But in the history of the Gum nebula we are confronted with the remarkable phenomenon of a vast celestial object that becomes visible only by mere chance.

The giant nebula did in fact exist even before the supernova explosion, but it could not be seen for lack of excitation. It became visible later on, a little at a time, as the radiation emitted by the supernova reached out farther and farther. It has been radiating since then, and it will continue to do so as long as its atoms recombine. However, lacking the source of excitation, which existed only for the few days of the outburst, the gas is no longer being re-excited, and therefore one day the nebula will vanish again.

Today we are able to observe that vast but tenuous object we call the Gum nebula. Prehistoric men could not have seen it even if they had had the most powerful telescopes; nor will it be visible to our progeny. In a way, it is like a landscape lit up in the dark of night by a flash of lightning. As far as the Gum nebula is concerned, we happen to live just at the moment of the flash—a flash, needless to say, that lasts many thousands of years.

SUPERNOVAE

The preceding paragraph began and closed with the hallucinating vision of an apocalyptic event. Although we cannot actually see it because of its dimness, and the low sensitivity to red of our eyes, the vast nebula looming over the southern hemisphere appeared in the sky only because a little over 10,000 years ago in that region of space there occurred a catastrophe that sowed death and destruction all around. Had there been planets like ours revolving about the exploding star, they would no longer exist today. Furthermore, the supernova explosion could have caused immense damage to any other planets in its environs, or at the very least to any form of life on them. By galactic standards the earth was relatively close to that exploding star, yet far enough not to suffer any serious consequences. Things would have gone quite differently for us, however, had the supernova been one of the nearer stars, or perhaps even the sun. It is quite reasonable therefore to worry about this possibility and wonder whether the same fate could indeed befall our sun.

CHARACTERISTICS OF SUPERNOVAE

To solve this problem we will try to find out which stars end in this manner, and whether any of them might have been like the sun before exploding. It is a very difficult question to answer. Contrary to what generally happens in the case of novae, we have never observed a supernova prior to its explosion. Furthermore, a supernova outburst is such a rare event that, since the time that we were in a position to study them with adequate tools, we have seen none relatively nearby, that is, within our own Galaxy. Only three galactic supernovae have been observed in the last thousand years—in 1054, 1572, and 1604. On the basis of ancient chronicles a fourth has been recently found which occurred in 1006 in the constellation Lupus.

All supernovae investigated with modern means from the time of maximum have occurred in external galaxies (figure 3.11). The first of these appeared in 1885 in the Andromeda galaxy (M 31) but was not properly investigated because nobody at the time quite realized yet the scope and significance of the phenomenon. In the following fifty years twenty more extragalactic supernovae were discovered by various observers. At the beginning of the 1930s W. Baade and F. Zwicky realized that these objects were quite different from novae and suggested that they should be called supernovae to distinguish them from simple novae. Zwicky then took this research a lot farther. Convinced of the necessity of gathering as many observational data as possible, he undertook a systematic search for supernovae and, for this purpose, had a small 46-cm Schmidt telescope built on Mount Palomar (one of the first instruments of its type). Starting in 1936, he carried out photographic surveys of celestial fields rich in galaxies in the hope that new supernovae would appear in some of them. His research was very successful, and nineteen additional supernovae were discovered in a little more than four years. After World War II the search was taken up again on an international basis and carried out in different observatories, including that of Asiago, Italy, beginning in 1959.

All these efforts contributed greatly to our knowledge of the phenomenon. By the end of April 1973 the number of known supernovae had risen to 378—and, of course, it keeps increasing. Furthermore, prompt discovery of a new supernova outburst enabled astronomers to follow its course with photometric and spectroscopic devices (the largest telescopes cannot contribute to the search because of their narrow fields of view).

At the conclusion of the first general symposium on January 1, 1941, several significant facts had emerged. First of all, the intensity of the phenomenon proved to be far more impressive than it has been previously believed. At the time the supernova reaches maximum, its luminosity nearly equals that of the entire galaxy in which it occurs, which generally contains a few billion stars as bright, on the average, as our sun. In addition, a first estimate of the frequency of the phenomenon was obtained by comparing the number of new supernovae with the number of galaxies under surveillance. It was estimated that a supernova appears in a galaxy, on the average, every three hundred years. Today the question of frequency of occurrence is somewhat more complex. Depending on the methods they used, various authors have found different values. It has also become apparent that the frequency depends on the type of supernova and on the mass of the galaxy in which it ap-

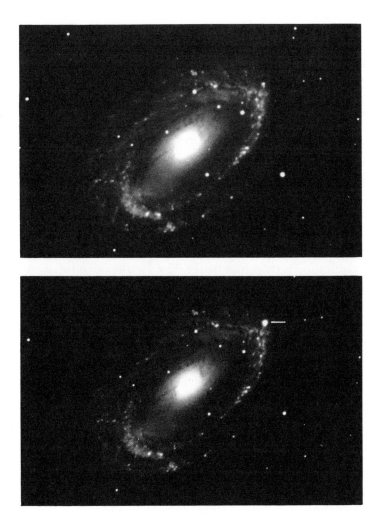

Figure 3.11   The galaxy NGC 4725 in Coma Berenices. Notice that the super-nova cannot be seen in the photograph made on May 10, 1940 (*above*), while it is quite evident in that of January 2, 1941 (*below*). (The Hale Observatories.)

pears. Consequently, we should define the classes of objects under discussion and refrain from sweeping conclusions.

In 1941 R. Minkowsky discovered that supernovae were not all alike. He found at least two types whose spectra and light curves were so different that they must be due to explosions in different kinds of stars.

In 1965 Zwicky improved this classification by adding three more groups. Type III and IV supernovae had been previously classified as type II. Many of these objects, however, were included only on the basis of their intense brightness at maximum and might not actually be true supernovae. Zwicky himself does not regard type III objects as being exploding stars but rather collapsing gas clouds. Type V includes atypical objects such as that in NGC 1058 and Eta Carinae, which, as we shall see in chapter 5, do not fit in with the normal picture of supernovae. The original classification is therefore better suited to the purpose of this book.

A distinction between type I and type II supernovae must start with their light curves. Figure 3.12 illustrates the main difference between the two types. The two curves were obtained by L. Rosino and his collaborators at the Asiago Observatory by combining the observations of 38 type I and 13 type II supernovae, respectively. Knowing the distance of the galaxy in which the supernova occurred and its apparent magnitude at maximum, one can compute its absolute magnitude. Using this method C. T. Kowal derived the absolute magnitude at maximum of 33 specially selected supernovae and found that it was nearly the same for all type I supernovae, with a value of -18.6. Type II supernovae are not as bright, attaining magnitudes of about -16.5. To put it simply, when these stars explode they become respectively 3 billion and half a billion times as luminous as the sun.[8]

The two types of supernovae can also be clearly distinguished by the different appearances of their spectra. Spectra of type II are similar to those of novae, the only difference being that a more pronounced

---

8. Note that adopting the most recent value of Hubble's constant ($H = 55$ km/s/Mpc), the absolute magnitudes of type I and type II supernovae increase to -19.9 and -17.8, respectively.

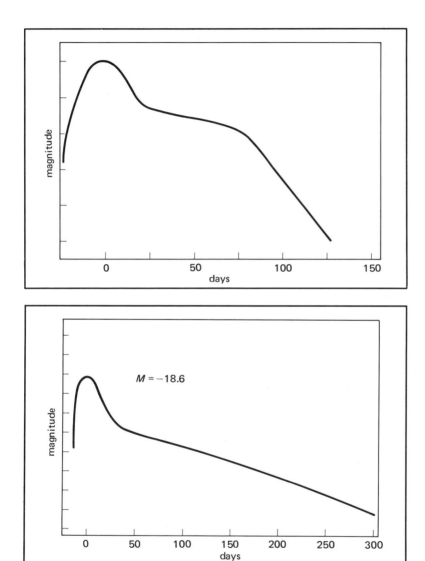

Figure 3.12   (a) The average light curve of 38 type I supernovae. (b) The average light curve of 13 type II supernovae. (From *Astronomy and Astrophysics.*)

broadening of the lines shows that the envelope formed by the explosion is expanding more rapidly. All spectra of type I supernovae show dark and bright bands that evolve with such regularity that the time of maximum can be determined just by examining a single spectrum obtained at any subsequent time. The nature of these bands has long been a mystery, but in recent years considerable progress has been made toward their identification with the help of instruments of greater resolving power and more refined theoretical interpretations. It has been found that a continuous spectrum makes a significant contribution to the luminosity. In addition, at least in some cases (notably for the H and K lines of ionized calcium), the emission peak, which is red-shifted due to the radial velocity of the associated galaxy, has been separated from the blue-shifted absorption. This enables us to compute the expansion velocity of the envelope.

In this manner it has been possible to calculate the maximum expansion velocities of the outermost layers of the ejected envelopes. They have been found to be about 20,000 km/s in type I supernovae and about 15,000 km/s in type II. Explosions of this kind are not even remotely comparable to any we are familiar with. We can barely begin to appreciate them if we realize that the above velocities are at least 250 times greater than that of the shock wave produced by a nuclear explosion in the immediate vicinity of the detonation point. And yet these are the incredible velocities with which supernovae expel matter into space, although somewhat lower in the inner shells.

The total mass of the ejecta has been estimated at about one-tenth the solar mass in type I supernovae and at about ten solar masses in type II. Type I supernovae are believed to originate from stars of about one solar mass, and type II from stars up to thirty times as massive as the sun. A comparison of the values found for the ejected masses clearly shows that the phenomenon is a great deal more impressive in type II than in type I supernovae, and certainly far more catastrophic than in novae, in which only a thin outer shell with a total mass one one-thousandth that of the sun is dispersed into space.

These findings are all extremely interesting, but they have not yet told us from which stars supernovae originate. An important clue comes from their distribution in the various kinds of galaxies. Whereas type I supernovae occur in all kinds of galaxies (spiral, elliptical, and irregular), type II supernovae are found in spiral but not in elliptical galaxies. In addition, they appear primarily in the outer regions of spiral galaxies, in or between the spiral arms. More precisely, J. Maza and S. van der Bergh have recently shown that type I supernovae appear over an entire spiral galaxy, while those of type II, which are much more frequent, are concentrated almost exclusively in the spiral arms. From these results it was deduced that type II supernovae are associated with young population I stars and originate from the evolution of massive stars, whereas type I supernovae are associated with old population II stars, and evolve from stars of about one solar mass.[9] This result has caused some problems to theoreticians: according to current theories on stellar evolution, at the present time normal population II stars more massive than about 0.8 $\mathcal{M}_\odot$ should no longer exist in our Galaxy. On the other hand, all four galactic supernovae that have occurred in the last one thousand years appear to be of type I, and therefore should have originated from population II stars about as massive as the sun. This difficulty could be overcome by assuming that the exploding star was not normal to begin with. For example, it might have been one of the two components of a binary system, where mass exchange from one star to the other alters the normal evolutionary process predicted by theory for single stars. We shall soon see that this possibility is not at all unlikely, at least for type II supernovae.

RUNAWAY STARS

In the constellation Auriga there is a variable star named AE Aurigae which is surrounded by a tenuous nebula. It seems that the star was not originally embedded in the nebulosity but "swept" it up about itself while passing through it (figure 3.13). The white arrow in figure 3.13 indicates the direction in which the star is traveling as well as the distance it covered in 50,000 years.

9. For a definition of populations I and II, and their evolutionary meaning, see *Beyond the Moon*, pp. 203–210.

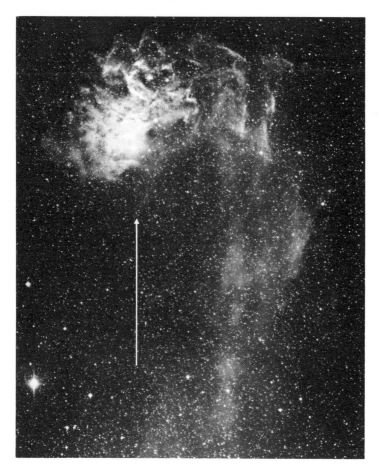

Figure 3.13   Motion of the variable star AE Aurigae. The star cannot be seen because it is embedded in the nebula. The arrow indicates the direction of motion and the distance traveled by the star in 50,000 years. (Palomar Sky Survey.)

If we retrace in time the path of this star, we find that on the basis of its velocity and direction of motion 2,700,000 years ago AE Aurigae was in the O association in Orion,[10] a region rich in stars of spectral type O where it would have been much more natural to find it, since AE Aurigae is also an O star (more precisely an O9).

This is not the only star that we have found outside its proper niche. We know two other stars, currently in different regions of the sky and far apart from each other, that appear to have originated from the same association in Orion. They are μ Columbae, a B0 star that is believed to have been in Orion 2,200,000 years ago, and 53 Arietis, a B2 star that apparently was in Orion 4,900,000 years ago (figure 3.14). In addition to the three originating from Orion, several other stars that are now scattered in the most diverse regions of the sky are known to have originated from other O associations, for example, those in Scorpius, Lacerta, and Cepheus. All these stars have two common characteristics: they are all very young hot stars of spectral types earlier than B5, and they are all moving through space at high velocities ranging from 50 to 150 km/s. This last fact is also very peculiar, since normal O and B stars usually move at low velocities of about 10 km/s. Both their high velocity and the direction of their trajectories indicate that these stars are not simply moving in accordance with the overall rotation of our Galaxy, but rather as if they had been expelled from their natural environs, that is from the O associations where other stars of the same spectral type can be found and where they themselves had apparently been born.

A. Blaauw called these celestial objects "runaway" stars, and he studied them extensively in the hope of finding the cause of their peculiar behavior. In the course of his research he discovered two interesting facts. First of all, stars of the heavier O and B types are ten times more common among runaways than normal O and B stars. In addition, none of the O and B runaways is a double system, whereas at least half of the normal O and B stars belong to binary or multiple systems.

As a result Blaauw formulated an intriguing hypothesis, which is ac-

10. For the definition and meaning of O associations, see *Beyond the Moon*, p. 180.

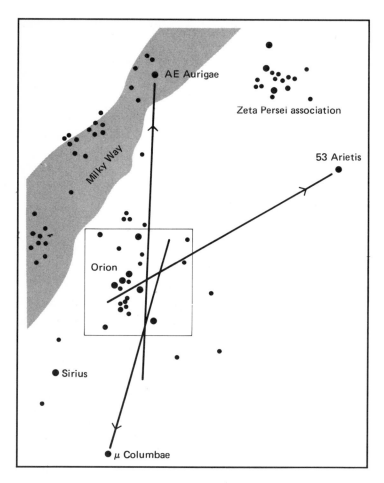

Figure 3.14   Tracks in the sky of three runaway stars originating in Orion: AE
Aurigae, 53 Arietis, and μ Columbae. The measured velocities are 106 km/s, 59
km/s, and 123 km/s, respectively. (From *Meyers Handbuch über das Weltall*, 1960.)

tually quite similar in its general outline to that independently proposed by Zwicky in 1957: O and B runaway stars once belonged to the O associations whence they in fact appear to have originated, and where they were born together with their companions. Originally, each of them belonged to a binary system and had a companion star that was very massive. Their companion evolved very rapidly, reached the supernova stage, and exploded. Following the abrupt destruction of the more massive component of the system, the remaining star could no longer move in a circular or elliptical orbit and suddenly flew off in a straight line along the tangent, with a slightly smaller velocity than it had in its orbit at that time. Let us assume for instance that the two stars had a mass of 250 and 25 solar masses, respectively, a relative distance of 20 AU, and an orbital period of 5.4 years. Assuming further that in the first three months of the explosion the larger star lost nine-tenths of its original mass, we find that the companion should have taken off at a speed of 90 km/s, which very nearly corresponds to that of many runaways. Naturally, we can also speculate that runaway stars with lower velocities originated from systems with less massive companions.

The high value postulated for the mass of the exploding star does not correspond to that normally found in binary systems. However, since we are dealing with systems of proto-stars, or newly formed stars, where the more massive companion evolved very rapidly, this fact does not compromise Blaauw's theory. If this had not been the case, the more massive star would not have had the time to become a supernova while the binary system was still in the O association. In fact, these associations are known to break up in a few million years, and only very massive stars can complete their evolutionary cycle in so short a time.

Interest in the theory that some stars would be flung into space like so many stones from a slingshot has been rekindled in recent years. The theory has been further developed particularly for the purpose of discovering whether binary systems involved in a supernova explosion must necessarily be disrupted. In certain close binary systems, for instance, there could be a mass exchange whereby the more massive and more rapidly evolving star gives most of its mass to its companion, so that, when it reaches the supernova stage and explodes, it no longer is the more massive component of the system. In this case the system is not disrupted, although substantial modifications will occur in the orbits of the two components as well as in the motion of the system in space.

The only difficulty with this interpretation appeared to be the fact that no pulsars had yet been found in binary systems. At the beginning of 1975, however, radio astronomers at the Arecibo Observatory discovered a pulsar in a binary system in which the companion was also an invisible collapsed object. Presumably we are dealing here with two neutron stars, one of which is also a pulsar. If true, that system would have withstood not one but two supernova explosions.

This field of research is therefore becoming increasingly interesting, and the theory of supernova explosions in binary systems is becoming more convincing and of more general application.

SUPERNOVA REMNANTS

It is evident that we have recently gained a great deal of information about supernovae; yet there are some essential points that are still unclear. For instance, we still do not know the distinctive characteristics of the exploding stars during the initial phases of the outburst or prior to it. This is not too surprising considering that we have never been able to observe a supernova before the explosion. What is surprising, instead, is that we have no clear understanding of what happens after the explosion, not even in the case of the nearest galactic supernovae where we can easily study what is left of the original star.

As mentioned at the beginning, and as generally understood, the evolution of the phenomenon appears by now quite clear. When the inner part of a star collapses, enough energy is suddenly generated to cause the outer layers to expand explosively into space. After a certain number of years these remnants appear like tenuous nebulosities, consisting both of the ejected material and pre-existing interstellar matter that was swept away by the shock wave and piled up along the wave front, much like snow pushed aside by a snowplow. What is left of the star is condensed into a very small and very dense object—a neutron star —which, while rotating at high speed, produces radio, optical, and X-ray pulsations and reveals itself to us as a pulsar. The region surrounding the explosion is filled with relativistic electrons, as shown by the polarization of the emitted radiation. These phenomena form a clear and complete picture which is perfectly understandable and in agreement with theory.

Unfortunately there is one serious problem: all this holds true for one object only—the Crab nebula. It was this supernova that provided most of the observational data used in the development of the theoretical model. Later, when new and unexpected facts about the Crab came to light, they were soon found to fit in perfectly with the theory, and in fact strengthened it. We have just begun to realize, however, that what has been found in the Crab nebula is not quite the same as what we observe in other remnants.

The first peculiarity resides in the visible characteristics of the various remnants. As we have seen earlier, the Crab nebula is fairly spectacular (figure 3.8). The same cannot be said for the two subsequent galactic supernovae, observed in 1572 and 1604, which have left barely discernible traces (figure 3.15). The remnants of the 1006 supernova, discovered by S. van der Bergh at the beginning of 1976, consist only of tenuous filaments.

Of course, we know other supernova remnants, some of which are quite conspicuous. Among the most beautiful are those in Vela (figure 3.9), the Cygnus Loop (figure 3.16), and IC 443 (figure 3.17). We have already discussed the first of these; the second is believed to have originated from a supernova that exploded about 60,000 years ago, and the third is from a more recent explosion. Associated with the visible remnants we find radio and often X-ray emission, due to collisional heating of the interstellar gas by the expanding ejecta. Naturally, the fact that these nebulosities are supernova remnants can only be deduced from theory and from their own expansion, since no supernova has been observed in the Galaxy other than those of 1054, 1572, and 1604. This is not surprising, considering that the outbursts must have occurred in prehistoric times, but there is a case that is quite disconcerting. A strong radio source called Cas A was discovered in 1948 in the constellation Cassiopeia. Subsequent optical observations revealed the presence of small knots and filaments, and later investigations disclosed also an X-ray

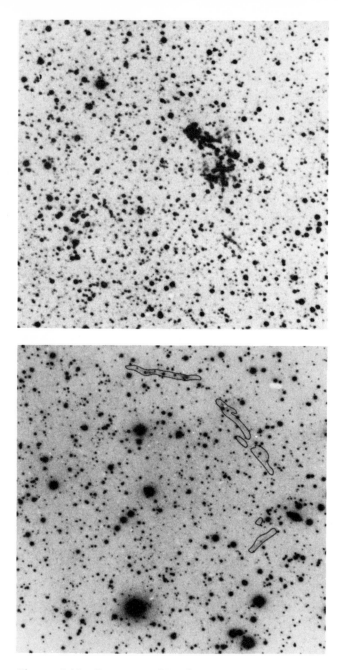

Figure 3.15 Remnants of Kepler's supernova of 1604 (*above*) and of Tycho's supernova of 1572 (*below*) photographed by S. van der Bergh with the 5-m telescope at Mount Palomar. The negatives of the photographs are reproduced (as in the original plates) because they show the very tenuous nebulosities better. Some of these nebulosities are delineated further in the bottom photograph by drawn-in lines. (Courtesy of S. van der Bergh.)

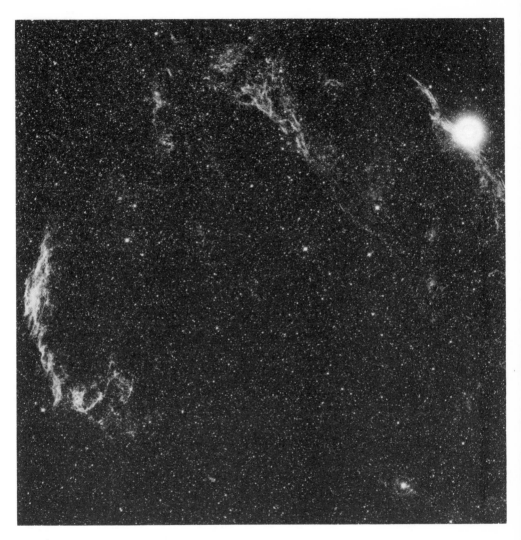

Figure 3.16   The Cygnus Loop, perhaps the remnant of a supernova, photographed in red light with the 65/90-cm telescope at Asiago Observatory. (Courtesy of L. Rosino.)

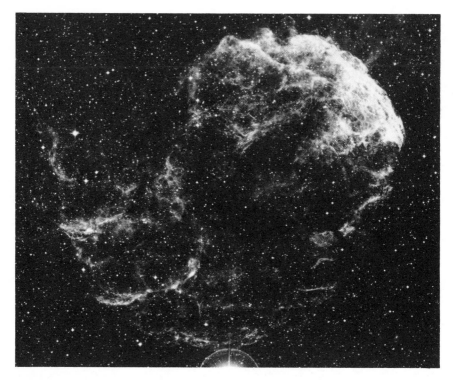

Figure 3.17   The gaseous nebula IC 443 in Gemini, a supernova remnant,
photographed in red light with the 122-cm Schmidt telescope at Mount Palomar.
(The Hale Observatories.)

emission. The optical and radio images are shown in figure 3.18, which also illustrates that a seemingly poor region can in fact be quite rich in material not optically conspicuous. From studies of the gas in Cas A we have learned that both knots and filaments are evolving very rapidly and, more importantly, that the gas itself was ejected by a supernova explosion in 1667. This date was calculated with an uncertainty of ±8 years. The supernova associated with these remnants should have been widely observed. But even though at that time several astronomers already kept the sky under surveillance, and modern means were already used to obtain, record, and transmit observations, there is no report of the event. How such an explosion could go unobserved is explained primarily by the hypothesis that a strong interstellar absorption in that direction may have so darkened the star that even at maximum brightness it remained an inconspicuous object. According to measurements made by L. Searle, however, the absorption is only of the order of 4.3 magnitudes. Consequently, if we adopt the distance of 2.800 parsecs found by S. van der Bergh and W. W. Dodd, and assign the supernova to type II with an absolute magnitude at maximum of −16.5, we find that the star should still have appeared as an object of magnitude 0, that is, as bright as Arcturus or Vega. Had it been a type I supernova, it should have attained an apparent magnitude at maximum of −2; in this case it would have been more luminous than Sirius, the brightest star in the sky.[11] In either case, it is certainly hard to understand how the observation was missed. Furthermore, if we assume a type II supernova, corresponding to lower brightness, the exploding star should have been very massive and presumably of class O. As we have seen, type II supernovae are believed to occur most often in O associations, but no stars of this type are known in the vicinity of Cas A.

Notwithstanding these difficulties, there is still some hope of explaining why the supernova was not observed. There could be a number of different reasons or a combination of them: the supernova may have

11. The situation would be even worse if we adopted the higher values for maximum brightness of supernovae based on the new value of Hubble's constant; see note 8.

Figure 3.18 The region of the radio source Cas A photographed in H$_\alpha$ light (*left*) and a computer map of the radio emission (*right*). In the second image the typical ring of supernova remnants can be clearly seen. (The Hale Observatories and Mullard Radio Astronomy Observatory.)

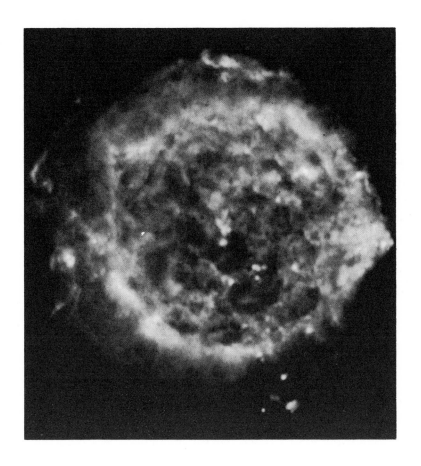

been slightly fainter or a little farther away; interstellar absorption may have been stronger; the maximum brightness of the star may have lasted for a short time only and may have been concurrent with a period of bad weather in Europe where most of the observers lived at the time.

There are other peculiarities that are more significant particularly because they apply to more than one case. Take, for instance, the fact that of twenty-five known supernova remnants only three appear to contain a pulsar: the Crab nebula, IC 443, and the Vela remnant at the center of the Gum nebula. Why is it that there are no pulsars in the other remnants? Perhaps only supernovae of a certain mass produce pulsars. Or it may be that not all pulsars can be observed. As we have seen earlier, their periodic pulsations are due to an intense rotating beam of radiation that produces a lighthouse effect. If the radiation was emitted only from one point and in a specific direction, the rotating beam would always lie in a plane, and we could observe it only in the rare event that the earth should be located in nearly the same plane. Furthermore, we cannot explain why the pulsar in Vela emits radio pulses without optical ones, while the Crab nebula emits both.

There is one last puzzling fact. Up to mid-1975, 147 pulsars had been observed and catalogued. As we have just seen, only three of them are associated with supernova remnants. If they all have the same origin, how can there be so many pulsars without remnants? A possible explanation is that many pulsars are so old that the remnants have completely disappeared in space. A point in favor of this interpretation is that most of these pulsars have periods greater than 1 second; as explained earlier, observations have shown that a pulsar's period becomes longer with the passage of time.

All this proves that there are still many unresolved questions and that we can not draw general conclusions from the observations of the 1054 supernova alone. Perhaps we will be able to find a solution to all these questions by determining at which "post-explosion" phase the object finds itself. It is also possible that there are different types of explosions, or at least different ways in which the remnants can evolve when one or more parameters in the exploding object vary (for instance, the mass) and, depending on the variations of the interstellar medium in which the remnants expand and evolve.

WILL THE SUN EXPLODE?

Now that we know more about supernovae, we can again ask the question, Could the sun explode? Undoubtedly, the answer is no. To begin with, the sun neither belongs to a binary system nor exhibits the spectrum characteristic of the first spectral classes. In addition, even if these conditions were not necessary, the sun is not massive enough to become a type II supernova, although it would have sufficient mass to explode as a type I supernova. This cannot happen, however, because supernovae of this type are population II stars, while the sun belongs to population I.

One can still worry that the sun might become a nova. Though much more modest on a cosmic scale, this event would be more than sufficient to sweep away all the inner planets. This will not happen either, since novae have very different characteristics from the sun, first and foremost being the fact that they all belong to binary systems. In fact, it appears certain that a nova explosion is caused precisely by the instability inherent in the system.

In this respect, therefore, we seem to be fairly safe. A danger could still exist, however, if not to the integrity of our planet, at least to the safety of mankind. Should a supernova occur relatively close to the solar system, the enormous amount of energy emitted by the explosion could be fatal to us on earth. Starting from this premise, and taking into account the energy emitted by a supernova as computed by several astrophysicists, K. D. Terry and W. H. Tucker estimated the probability of the earth receiving a lethal dose (about 1,000 r) of cosmic rays and X rays. The conclusion reached by the two Americans is that such an event might occur at least once in 150 million years.[12] This estimate is based on the hypothesis that one type II supernova explodes in the galaxy every fifty years. According to the most recent estimates, however, these events appear to be less frequent. If so, nearby explosions would be more infrequent as well, and the chance of our being struck by a lethal

12. See *Beyond the Moon*, p. 165.

dose of radiation would become much smaller. However, as we said, part of the dangerous radiation consists of X rays. Until a few years ago we had no way of measuring X-ray emission from supernovae, but today the use of X-ray telescopes carried aloft by artificial satellites enables us to observe the sky unimpeded by the earth's atmosphere. A type I supernova appeared in 1972 in the galaxy NGC 5253, which is closer and therefore easier to study than the galaxies where supernovae are usually detected. The satellite OSO-7 scanned that region for X-ray emission from ten days before the explosion to one hundred and fifty days after; other observations of the region spanned an even longer period, from sixteen months before the optical maximum to two years after. No X-ray flux was ever detected. If one was present, it could not exceed 2×10^{-10} erg/cm^{2}s. Thus it was concluded that in the 1 to 67 KeV range the supernova's emitted power did not exceed or equal 3×10^{42} erg/s.

From our point of view this result is very reassuring. Unfortunately, nothing is known as yet about X-ray emission from type II supernovae, which in this respect are the more interesting. Furthermore, there might be invisible exploding stars emitting only, or almost only, X rays.

Recently, even this event has been observed.

TRANSIENT X-RAY SOURCES
In 1969 two Vela 5 satellites[13] detected the sudden onset of an X-ray source in the southern sky. A similar event had already been observed to occur in Cen X-2, which had suddenly appeared two years earlier. On that occasion it was not possible to determine if the onset had been gradual or abrupt, since prior to discovery the region had not been scanned for seventeen months. The 1969 source appeared between 11:30 P.M. on July 6 and 4:30 A.M. on July 9. In a period of fifty-three hours, or possibly less, the X-ray emission increased from a level below the minimum detectable to an intensity twice that of Sco X-1, the brightest X-ray source in the sky except for the sun.[14] Past maximum the

13. W. D. Evans, R. D. Bellian, and J. P. Connors, 1969.—Trans.
14. Sco X-1 was the first X-ray star discovered in the night sky by Giacconi, H. Gursky, F. Paolini, and B. Rossi in 1962.—Trans.

emission began to decrease in intensity, its spectrum shifting toward softer X rays. This behavior was similar to that observed in the decreasing phase of Cen X-2.

The new source, designated as Cen X-4, did not remain long an isolated case. In 1971 another source of this type was observed in the constellation Lupus by the Uhuru satellite. It was first detected on August 17 as a weak X-ray source of intensity approximately one-tenth that of Sco X-1. On August 23 it reached an intensity twice that of the X-ray source in the Crab nebula. It then began to fade, and by December 20 it had decreased to one-tenth of this value. The sudden and rapid increase followed by a slow decay to a minimum was reminiscent of the optical light curves of novae. A search for an optical counterpart on plates obtained at the Harvard and Boyden Observatories was unsuccessful. No novae brighter than magnitude 15 had occurred in that region.

However, phenomena of this type continued to be observed, and by 1974 nine cases had been studied. The average light curve of the X-ray emission derived from these data (figure 3.19) shows great similarity to the optical light curve typical of novae. The possible relationship with novae had been suggested earlier by J. L. Elliot and W. Liller, who in 1972 had pointed out the similarity of the X-ray emission profile in two transient sources (2U 1543-47 and Cen X-4)[15] with the optical emission profile of the novae T Pyx, DQ Her, and IM Nor. In addition, the same authors suggested that IM Nor may even be coincident with one of the transient X-ray sources (2U 1536–52).

In 1975 there was an unexpected breakthrough in this field when the Ariel satellite, fifth in the British series, discovered and observed at length the most interesting and best studied of these X-ray novae. The satellite had been launched from the San Marco platform in October

15. X-ray sources are generally designated by a letter (the initial of the satellite from which they have been discovered) followed by four numbers (right ascension, in hours and minutes), and after a + or − sign, by two more numbers (declination). Other designations are given by the name of the constellation in which the source is found, followed by the letter X and a number indicating the order of discovery.

1974 and had already observed a number of sources of this type. The exceptional source was found on August 3, 1975, by the University of Leicester astronomy group,[16] and received the name A 0620-00. At the time the source was relatively weak, but in subsequent days it kept increasing in intensity until on August 14 it was three times brighter than Sco X-1 (figure 3.20). At the time of discovery the emission was prevalently observed between 2 and 18 KeV, but on August 7, while increasing toward maximum, the emission abruptly shifted below 10 KeV. Apparently, this shift marked a change in the energy production mechanism, which caused a transition from the emission of relatively hard to relatively soft X rays. Starting from mid-August the intensity of the source began to decrease according to the well-known pattern.

In the meanwhile optical searches were under way, and on this occasion they were crowned with success. On August 15 two Dartmouth astronomers using the Kitt Peak 132-cm telescope found a 12th magnitude star at the position of the X-ray source. (In a 1955 plate of the Palomar Survey the star appeared extremely faint—magnitude 20.5.) On the following nights low-dispersion spectra were obtained which appeared completely devoid of either emission or absorption lines. Further observations, some of which were carried out with the 5-m Mount Palomar telescope, confirmed that the spectrum was dominated by the continuum. Superimposed on the continuum, a few absorption lines or bands due to matter between us and the object could be observed, as well as a line of He II and some of the hydrogen lines in the Balmer series, which were in emission and very weak.

The object was also detected as a radio source, and its emission was monitored by several radio telescopes. After August 18 the intensity in radio also began to fade, and henceforth it continued to decrease more rapidly than the X-ray emission. These observations, and particularly the fact that they had started during the rise many days before maximum, gave us for the first time a comprehensive view of the phenomenon.

But the most startling discovery came from optical observations.

16. K. A. Pounds, M. Elvis, C. G. Page, M. J. Ricketts, and M. J. L. Turner—Trans.

Searching old plates of the region in the Harvard collection, it was found that in November 1917 the 20.5 magnitude star had exploded once before in visible light, attaining photographic magnitude 12 on that occasion as well. Therefore there was good reason to conclude that the star was a nova, and moreover a recurrent nova. Even the rise in intensity of 8.5 magnitudes fitted in well with the characteristics of these objects, and in fact allowed astronomers to specify an average cycle of half a century, which closely corresponds to the interval between the observed outbursts.

In spite of these findings, the hypothesis that we are dealing with a nova with a strong X-ray emission is not consistent with our knowledge of novae, for instance, with regard to their spectra. On the other hand, several theoreticians have already interpreted the behavior of A 0620-00 in a completely different way. In their opinion, the observed phenomena could be explained in terms of binary systems in which one of the components is an X-ray source. In these binary systems one of the two stars exploded as a supernova, but the system remained bound as in the special case discussed in the paragraph on runaway stars. In view of this the occurrence of an X-ray maximum would be due to orbital motion, and so should be periodical in nature. However, this interpretation is still controversial, and does not explain all the mystery still attached to this phenomenon.

Given the small number of X-ray novae observed, none of the current theories can be considered sufficiently well founded. As a result we will not discuss any of these in depth. But the reader may be interested to learn that in addition to the known explosions that we have observed visually in the last two millennia, there are others that are undetectable in visible light but no less impressive. We had just concluded that perhaps the fear of supernovae was without foundation, and now we find that the same danger could come from a new quarter. From what we have learned up to now it does not seem that the radiation doses produced by X-ray novae are dangerous, but the very existence of these objects poses a new question: Do we really know the denizens of space well enough to be sure that we can never be hit by a death ray?

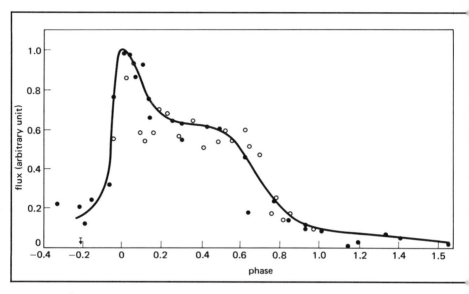

Figure 3.19   Average light curve of X-ray-emitting novae. The experimental data from Cen X-4 are shown with circles, and those from eight other novae are shown with dots. (From *Astrophysics and Space Science.*)

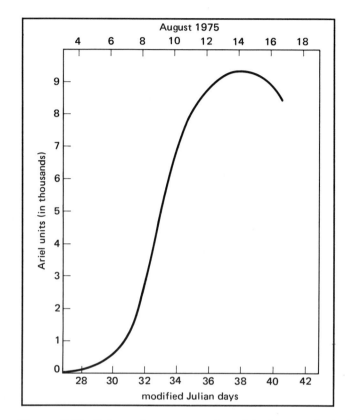

Figure 3.20   Schematic representation of the increase in the X-ray emission from A0620-00 around maximum, which occurred on August 14, 1975. (From *Nature*.)

SUPER-EXPLOSIONS

THE END OF THE DINOSAURS

Some years ago while conducting aerial surveys of the southern Sahara desert, a group of French uranium prospectors stumbled upon a sensational discovery: in the region of the Gaudoufouà, north of Agadés oasis, they found a dinosaurs' graveyard. All who have seen it have been struck by the feeling that here was one of the greatest discoveries about the history of the earth—direct evidence of one of the most awesome biologic catastrophies of all times. Skeletons of prehistoric animals, like those painstakingly reconstructed by paleontologists with pieces from different specimens often found in different places, emerge here out of the desert sands in countless numbers, while many others certainly lie buried and hidden. They are the petrified remains of huge beasts that seem to have died only yesterday but in fact lived in this region millions of years ago where in place of dry burning sands there were forests and streams.

They were enormous animals, often 20 to 30 m long, as tall as houses and weighing as much as 100 tons. Although they are known by the generic name of dinosaurs, they belonged in fact to a vast class of reptiles that included a wide variety of types. Some resembled giant lizards that would slither rather than walk over the land, and were often to be found resting in swamps and rivers where the water helped them carry their weight. Others were flying reptiles that differed from our largest and most ferocious birds, as ostriches from hummingbirds.

Despite their monstrous size, they were generally meek and peaceful creatures. They were mostly herbivores, even though among them prowled some ferocious killers, such as Tyrannosaurus, a 10-m tall, 6- or 7-ton monster and certainly the largest, most vicious carnivore that ever walked the earth. All these animals, and others with them, once lived, reproduced, and died in what is today one of the most forbidding deserts but was then a hospitable land capable of supporting life.

All of a sudden and in a very mysterious fashion these huge animals became extinct—not only those living around the swamps and rivers of what is today the Great Teneré Desert but all dinosaurs all over the world. They had appeared on earth at the end of the Jurassic Period, 140 million years ago, lived unopposed for about 80 million years, and became extinct between 60 and 65 million years ago. Their end was very quick. Until recently it was believed to have taken place within the span of about half a million years, but today we can narrow it down to no more than 10,000 years. This is an extremely short time as compared to the length of their existence on earth. Note also that with today's devices we cannot date remains with a finer accuracy than 10,000 years; so we cannot exclude the possibility that they could have disappeared in a much shorter time, perhaps in a matter of years or even days.

Scientists have long tried to solve this extraordinary enigma with various explanations and theories, but none of the proposed solutions is entirely satisfactory. Even theories that at first appeared quite well founded have ultimately failed to explain the two most puzzling features of the event—its worldwide diffusion and the rapidity with which it happened.

Both would be satisfactorily explained if we were to assume a cosmic source of destruction: in dealing with a source outside the earth, a lethal radiation could envelop the entire planet at once. Its rapid effects would be due to concentrated emissions in a period of a few days or months. We have seen earlier how Terry and Tucker suggested a supernova as the possible cause, and how doubts were cast on their theory when no X-ray emission was found to come from the supernova in NGC 5253. However, we have also learned that there might be intense X-ray outbursts with very faint optical counterparts. Whereas in either case evidence is inconclusive, we would do well to consider the possibility that there might exist different types of X-ray outbursts and sources. Recently, this hypothesis has received support from observational evidence.

In 1972 two Canadian astrophysicists, V. A. Hughes and D. Rutledge, confirmed that the solar system is near the center of an enormous gas ring that is rapidly expanding and dispersing into space. (Two similar rings had already been discovered in the Large Magellanic Cloud, one of the two nearest galaxies.) In 1967 P. O. Lindblad had observed in the 21-cm line a neutral hydrogen band that in view of its longitudinal extension had to be relatively close to us. This configuration could have been a ring similar to those observed in the Magellanic Cloud, with the solar system, and therefore the earth, inside of it. In this case the hydrogen belt should have been observed to describe a complete circle around the sky. Detection of this belt was very difficult, however, particularly toward the galactic center which is itself a strong radio source. The two Canadian astrophysicists attempted to reconstruct it by observing the hydrogen in absorption against the background of galactic emission. They were thus able to complete the circle and confirm that Lindblad's band was in fact a ring circling the entire sky. Rather than a true circle, they actually found it to be a large ellipse whose center is 900 light-years from the sun and whose major axis is about 4,200 light-years in length. This elliptical belt appears to have an enormous mass, 3 million times greater than that of the sun, and to be expanding at a rate of 6 km/s.

We should also call attention to the 1962 work of J. D. Fernie on the distribution of interstellar dust in the vicinity of the solar system. According to this author, the dust appears to be distributed in a shell whose dimensions are more or less those of the expanding ring. The observed dust could therefore be a residue of the explosion.

Even more significant is a comparison of the neutral hydrogen ring with the so-called "Gould's belt." The existence of this chain of stars in the vicinity of the solar system was pointed out by B. A. Gould between 1874 and 1879, but his work was later criticized. The most recent studies in this field (J. R. Lesh, 1972, P. O. Lindblad et al., 1973, R. Stothers and J. A. Frogel, 1974, and O. J. Eggen, 1975) have not only confirmed the existence of this group of stars but described its characteristic features with increasing accuracy.

Here is a brief description of it. In the solar neighborhood there exists a dense aggregate of stars known as the local system, which is flattened in shape like the Galaxy, but with a thickness of only 600 light-years, and whose equatorial plane has an inclination of 20° to the galactic equatorial plane. This vast conglomeration of stars has a diameter of about 3,200 light-years. Since we are near the system's equatorial plane, we see the component stars mostly projected on a broad ring that makes a complete circle around the sky and has a thickness equivalent to that of the system. Gould's belt, as this ring is known, comprises a great number of stars of spectral type O and B. The scientists investigating this phenomenon have found that all the stars in the belt are very young, though not all of the same age. Those in the central region appear to be 30 million years old; the age of the α Persei association, which is part of the belt, is between 50 and 60 million years, while that of the single component star ranges from 10 to 50 million years. Like the ring of neutral atomic hydrogen observed by Hughes and Rutledge, this belt is expanding at a rate of 10 km/s, as computed by Lesch.

Three very significant facts have emerged from all these investigations: (1) there is a gas ring that very nearly coincides with a belt of young stars, although the latter appears somewhat smaller in size; (2) the solar system is located not too far from their common center; (3) the two rings are currently expanding. Various theories have been proposed to explain this expansion. H. Weaver envisages the interaction of two gas streams originating from the northern and southern galactic hemispheres, respectively, and meeting on the equatorial plane in the region from which the gas appears to be streaming. Hughes and Rutledge suggest instead that in a very remote epoch a tremendous explosion must have occurred in the central region, of which the ring of neutral hydrogen, the interstellar matter, and the young stars in Gould's belt are either the remnants or the secondary products. From the current dimensions of the rings, and their computed expansion velocities, the explosion could be traced back to its origin. The result obtained by the two scientists is a startling 65 million years ago, which is exactly the time of the dinosaurs' mass extinction.

This coincidence is truly astonishing not only because the time of the explosion corresponds to that of the dinosaurs' extinction but also because the age of the stars in Gould's belt is always smaller than, or at most equal to, that of the hydrogen ring. This means that all the stars in the belt appear to have formed after the explosion. This explosion must have been truly awesome. But even though it occurred at a greater distance from us than that considered dangerous by Terry and Tucker, it is still possible that the amount of energy released was large enough to be fatal to animals on earth. We can conclude from a computation of the energy released in the event and from the very large mass of the remnants that the explosion was much more powerful than that of the usual supernova. What kind of supernova it was we do not know, and we cannot exclude the possibility that it might not have been a supernova at all. What appears certain is that such a violent explosion was much too close. A little less than a thousand years later, when most of its power would still be concentrated, great amounts of radiation could have reached the earth and traveled on. The ring we are now observing would thus be the diluted, trailing edge of a lethal wave that reached and devastated the earth, and then went on to dissipate into space. Without ruffling the seas or moving a leaf, this invisible hurricane sowed death everywhere on the planet. At that time, fortunately, man had not yet been born.

SUPER-SUPERNOVAE

As already mentioned, a ring similar to that discovered in our Galaxy, not too far from the solar system, had been observed in the Large Magellanic Cloud, one of the two irregular galaxies associated with our own Milky Way. The region had been defined by H. Shapley, and he classified it as constellation III in his morphological description of that galaxy. In 1966, while investigating supernova remnants in the Cloud, B. E. Westerlund and D. S. Mathewson pointed out some very puzzling facts. Constellation III is peculiarly shaped as an ellipse whose lower portion is marked by a chain of very young blue stars, while the upper part is defined by an arc of neutral hydrogen, discovered by radiotelescopes and invisible in the photograph (figure 3.21, upper left-hand corner). Two remnants of type II supernovae (R_1 and R_2) have been found just in that part of the elliptical ring that corresponds to the hydrogen arc. The region within the ellipse is strangely devoid of hydrogen, and contrary to what we generally observe in associations, even the group of young blue stars near the center (A) is not embedded in neutral hydrogen.

The entire ellipse is more than 3,000 light-years across, and the envelope, consisting of stars and neutral hydrogen, has a mass equivalent to 35 million solar masses. Westerlund and Mathewson have suggested the possibility that this whole region might be the product of a super-explosion of the type suggested in 1960 by the Russian astrophysicist

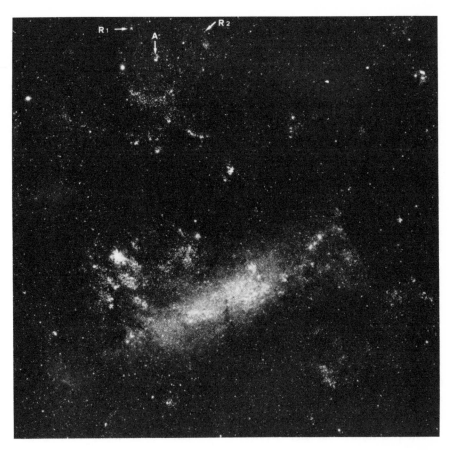

Figure 3.21   The Large Magellanic Cloud, with Constellation III (around point *A*) and the two supernova remnants (R_1 and R_2). (Mount Stromlo Observatory.)

I. S. Shklovsky. If the material ejected by the exploding object had been hurled into space at a speed of 5,000 km/s, it should have had a mass of the order of 100,000 $\mathcal{M}_\odot$, and would have kept on expanding for at least 3 million years.

A few years earlier, similar rings had been discovered in other galaxies by R. Hayward. Furthermore, as previously mentioned, Westerlund and Mathewson had noted the presence of two supernova remnants at the edge of the possible residue of a much greater explosion in the Large Magellanic Cloud. Similarly, two other supernovae specialists, C. T. Kowal and P. Wild, found that the supernova discovered in NGC 4189 in 1966 was located at the edge of a sort of round hole which unaccountably interrupted the outermost spiral arm of that galaxy. Wild then re-examined the positions of all supernovae on the best photographic plates of the respective galaxies and found five additional cases of this type. This confirmed the suspicion that, at least in some instances, supernovae may be set off by explosive events of far greater scope.

The known cases are too few and too unclear for us to draw general conclusions. Nonetheless, the two discovered in our Galaxy and in the Large Magellanic Cloud, respectively, suffice to show that far larger explosions than supernovae have almost certainly occurred in the past, and might still happen in the future. They certainly seem to provide strong support for the super-supernova theory proposed by Shklovsky.

The possibility that such events might occur had been suggested in an attempt to explain the formation in the universe of the heavier elements, from iron on. During the initial phases of galactic evolution the primordial gas consisted almost exclusively of hydrogen. From this gas was born a first generation of stars (population II), within which there formed the lighter elements, such as helium, carbon, oxygen. There is, however, no known group of reactions that could lead to the formation of the heavier elements in normal stars. According to theoretical calculations, this process could instead occur in supernovae, which by their own explosion also provide the mechanism for injecting the new material into space. Second generation stars, or population I stars, such as our sun, formed from this new material which also contains metals. Had these explosions always been limited to stars of small mass, from which type I supernovae currently originate, and had they occurred with the frequency we observe today, the percentage of metals should be much lower than what we actually see. According to Shklovsky, however, the type I supernovae that exploded in the initial stages of galactic formation were enormously more powerful and frequent. These super-supernovae (some of which might still go off!) are believed to manufacture and inject into the surrounding space a wealth of material consisting also of the heavier elements. This material, distributed in a ring, is further enriched by pre-existing matter swept away by the tremendous shock wave. From this material new stars are born, as shown by Gould's belt which coincides with a hydrogen ring and consists of a great number of stars whose ages do not exceed 60 million years. Some of these second generation stars, heavier than others, may evolve more rapidly and become type II supernovae. The two objects discovered on the ring in the Large Magellanic Cloud would be the remnants of just such supernovae.

From this new point of view supernovae do not appear only as the killer monsters we believed them to be at the beginning; they are also the seed of a new phase of cosmic evolution which heralds the coming of an environment closer to the one we know. After the supernovae's self-destruction, new stars are born from the remnants which will evolve and perhaps one day have planets capable of supporting life.

Once again we have thus come to realize that reality has many contrasting facets depending on the points of view. In a stormy night the sea appears as an uncontrollable force of nature that relentlessly seeks to draw us into the deep; yet, when sunlight shimmers on its gentle swells, its beauty fills our hearts, and we remember that this vast world nurtures in its briny depth creatures beyond count. Similarly, if we regard these awesome explosions in space from a narrow and self-centered point of view, they appear only as harbingers of death and destruction, not only to the worlds closest to them but also to the inhabitants of remote planets. Viewed from afar in their age-long evolution, they become part of a complex and wondrous process in which violence begets harmony and death brings forth life. Thus, once again, beginning and end appear inextricably intertwined, and the meaning of this process—if any exists —we know not.

4 ETA CARINAE

AN EXTRAORDINARY OBJECT

HISTORY AND INTERPRETATIONS

In a region of the southern Milky Way, which is a veritable jewelbox of blue and light blue stars, there is an extended nebula intersected here and there by dark lanes that are known to be dust clouds placed between us and the nebula. Figures 4.1 and 4.2 show wide and very wide-field photographic plates of the region. Unfortunately, neither of these photographs conveys the thrill of observing directly through the telescope what is one of the most spectacular sights in the sky. In the lower left-hand corner of figure 4.2 two lines have been drawn that, when extended, will meet at a star embedded in the nebula. This star is designated as Eta of the southern constellation Navis or, more accurately, Carina. Though barely visible to the unaided eye, it is undoubtedly the most extraordinary and puzzling star in the sky—if star is the right word for it.

The constellation it belongs to—Argo Navis—is very ancient, almost certainly dedicated to the memory of the Argonauts who sailed in quest of the Golden Fleece. Because of its vast size (it is the largest in the sky), around the middle of the eighteenth century it was divided by N. L. La Caille into three parts: Carina, Puppis, and Vela.[1] For this reason our star is known today as Eta Carinae, but is designated in many old atlases as Eta Argus. Its 1950 coordinates are: R.A. = $10^h43^m07^s$; Dec. = $-59°25'16''$.

As far as we can ascertain the astronomers of the Hellenistic period did not know this star. Ptolemy did not mention it in his *Almagest,* although he had observed the brightest stars in Navis; nor did F. V. Houtmann notice it in 1600 while he was examining all the stars in that region up to magnitude 4. Its name appears for the first time three years later in the 1603 celestial atlas of J. Bayer, where it is listed as a star of the 4th magnitude. E. Halley, who observed it from the island of St. Helen in 1677, also recorded it as a star of the 4th magnitude, but nearly a century later La Caille found it brighter (m 2).

---

1. Keel, stern, and sails, respectively.—Trans.

From 1811 Eta Carinae was kept under observation for many years by J. W. Burchell, a botanist who was also an excellent observer of celestial phenomena. During his stay in Africa, which lasted until 1815, he observed it several times and always judged it to be of the 4th magnitude. In December 1829 Burchell resumed his observations of the southern sky from São Paulo, Brazil, and to his great amazement discovered that Eta Carinae had become a star of the 1st magnitude, as bright, that is, as the brightest star in the Southern Cross.

By February 1830 the star had faded slightly by about one magnitude. From 1834 to 1837 John Herschel, son of the famous Wilhelm Herschel, observed it from the Cape of Good Hope, where he had gone to carry out observations of the southern sky. He always found it to be between the 1st and 2nd magnitude. In December 1837, while carrying out photometric measurements of the stars in that region, he noticed that Eta Carinae had brightened again, and in the following month it was brilliant enough to rival the brightest stars in the sky. Then it faded once more by about one magnitude. In March 1843, when John Herschel was back in the northern hemisphere, Eta Carinae attained a luminosity that had never been observed before, outshining even Canopus (-0.9 magnitude). During this period, which apparently lasted for a few months, its orange-red light dominated the southern sky and changed the appearance of the whole region. At that point only Sirius, the brightest star in the sky (m 1.6), still outshone the brilliant object that a century and a half earlier had not even been visible to the naked eye. Sirius's greater luminosity was only apparent. At a distance of only 8.8 light-years, it is one of the nearest stars, while Eta Carinae, as we shall see later, is about 8,000 light-years away from us. Had they both been at the same standard distance of 33 light-years, their apparent visual magnitudes would have become 1.4 and -13.0, respectively. Thus in the spring of 1843 Eta Carinae was actually almost 600,000 times brighter than Sirius, and more than 12 million times brighter than the sun. Had it really been at a distance of 33 light-years, which is the standard for absolute magnitudes, it would have lit up the landscape at night more intensely than the full moon. Such brightness is exceeded only by that of a supernova in the few days it is at maximum. We can confidently say that in the spring of 1843

Figure 4.1   The region surrounding Eta Carinae (invisible at the center) is shown in a wide-field photograph. The most luminous region is increasingly enlarged in figures 4.2, 4.7, 4.8, and 4.9. (From the *Atlas of the Southern Milky Way*; courtesy of H. Haffner.)

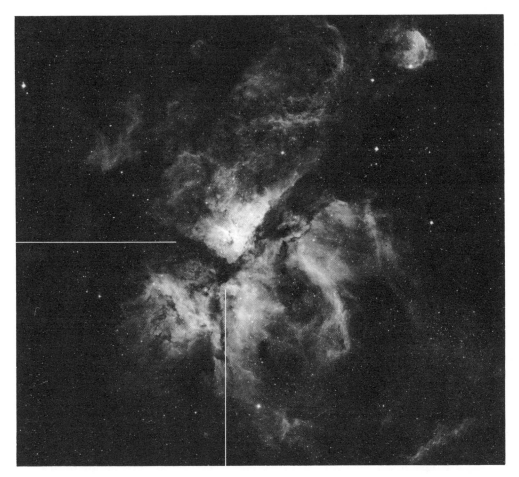

Figure 4.2 The Carina nebula photographed by B. J. Bok with the Schmidt telescope at Cerro Tololo Inter-American Observatory in Chile. The field is a much enlarged version of figure 4.1. The star is located at the intersection of the two drawn lines. (Courtesy of B. J. Bok.)

Eta Carinae was the brightest star in the entire Galaxy, which means the brightest of more than 100 billion stars.

In subsequent years its brightness gradually diminished, except for secondary peaks in 1856, 1871, and 1889, and it eventually stabilized around the 8th magnitude. Since 1940, however, it has begun to brighten again, and in 1976 reached the threshold of naked-eye visibility.

The variations in brightness exhibited by Eta Carinae in the last 300 years might correspond to a cyclical process. In fact, as suggested by the Assyrologist Peter C. A. Jensen, Eta Carinae might even be the brilliant and mysterious star that according to Babylonian inscriptions was already subject to occasional dimming a few thousand years ago.

Whatever its place in ancient history may be, the only reliable observations we have are those carried out from Halley's time to the present. The light curve based on these measurements is shown in figure 4.3.

The extraordinary fluctuations in luminosity exhibited by Eta Carinae have been accompanied by equally exceptional spectroscopic phenomena. Unfortunately, no data are available for this, the most interesting period of maximum brightness, nor for the period prior to it. The first visual spectroscopic observation was carried out by A. Le Sueur in 1869. It was not until three years later that astronomers began to record photographically the spectra of the brightest stars, such as Vega. From Le Sueur's description, the spectrum does not appear to have been very different from the current one. Twenty years later the first spectrographs of Eta Carinae were obtained at the southern station of the Harvard Observatory, and they showed the typical features of supergiants of class F5.

These spectra, and particularly the subsequent ones, revealed the most astonishing collection of emission lines ever observed in the spectrum of a star. Figure 4.4 shows portions of spectra obtained at the Cordoba Observatory between 1944 and 1959. What is most remarkable about them is the great number and intensity of the emission lines. One immediately notes the Balmer series of hydrogen, in particular the red H_α line responsible for the star's characteristic color. A more detailed study of these spectra has disclosed other significant facts. In addition to being intense, the hydrogen lines are also quite complicated; in particular, they show the P Cygni effect, that is, an absorption line shifted to the violet of the emission line. This effect, which we will discuss in connection with P Cygni, reveals the presence of an expanding shell. Furthermore, the few absorption lines that cross the weak continuum also appear to be shifted toward the violet. From the amount of their displacement with respect to the emission lines, one can determine that the absorption lines are produced by a shell expanding at a speed of 475 km/s. Through research carried out mostly before 1953, quite a number of emission lines have been identified. In addition to the Balmer series, one finds lines of [S II], Ti II, V II, Cr II, [Ni II], and, more important, lines of Fe II, both permitted and forbidden. Particularly significant was the discovery of the forbidden lines of [Fe II], (P. W. Merrill, 1928). Their presence is somewhat mysterious. As a rule they appear only in a specific period of the evolution of slow novae, that is, when they are fading back to minimum, but in the spectrum of Eta Carinae they are always present. The excitation level corresponding to these lines is not very high, and indicates a temperature for the star of 8,000°K. This value fits in well with spectral class F5, to which Eta Carinae is believed to belong on the basis of the continuous spectrum and absorption lines. On the other hand, emission lines of He I, [N II], [Ne III], and [Fe III], corresponding to a higher excitation energy, have also been identified. This can be explained if we assume that the various lines are formed in shells located at different distances from the center, as in planetary nebulae but with the difference that in this case the shells are much closer to the central object and to each other. For this reason, and also because of the star's great distance, everything is reduced to a point and we cannot separate the various shells in photographs taken in different lights.

To place the extraordinary characteristics we have discussed so far in better perspective, we must take into consideration the distance of the object. As we noted, when we derived the absolute magnitude at maximum, it is only by taking into account its distance that we can determine the intrinsic properties of the star; and by comparing these properties with those of other known celestial objects we can better gauge

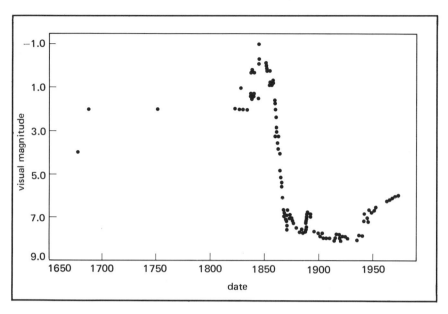

Figure 4.3 The light curve of Eta Carinae from 1677 to the present. (From *Astronomy and Astrophysics*.)

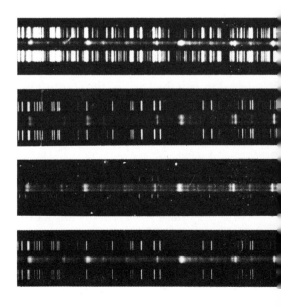

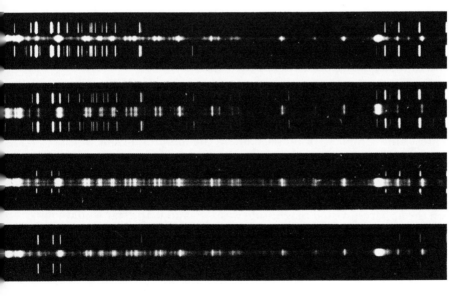

Figure 4.4   Four spectra of Eta Carinae obtained in 1944, 1951, 1955, and
1966 (top to bottom) at the Bosque Alegre Observatory in Cordoba. Comparative laboratory spectra are shown above and below each spectrum. Notice
that the intensity of several emission lines changes in time with respect to the
continuum.

to what degree the star is out of the ordinary. Unfortunately, it is very difficult to measure the distance: Eta Carinae is too far away for us to use the method of trigonometric parallaxes and is so atypical that all other methods based on absolute magnitudes, such as that of spectroscopic parallaxes,[2] cannot be properly applied. We will discuss later the method most commonly used. Right now we will only mention that the latest results place Eta Carinae at a distance from earth of about 7,500 light-years—or, at any rate, of no more than 8,200 and no less than 4,900.

In the two extreme cases, we find that the absolute magnitude at maximum was −13 for the maximum value of the distance, and no less than −11.9 for the minimum. Considering that the apparent magnitude at maximum was −1 and the current one is about +6, we find an excursion of 7 magnitudes. This means that even now, at near minimum, Eta Carinae's absolute magnitude is −4.9 (for minimum distance), or −6.0 (for maximum distance). Thus even past the climactic period Eta Carinae is still one of the brightest stars we know, much like the O supergiants of the first spectral types, which are, however, short-lived on account of their prodigious expenditure of energy.

This result becomes even more impressive when we compute the amount of energy that the star emitted just in the 250 years it was brighter than 4th magnitude. L. Gratton performed this calculation adopting the most unfavorable hypothesis, namely, the minimum value of the distance and assuming zero bolometric correction (or that the bulk of the star's energy had been radiated in the form of light). Under these conditions he found $M_{bol} = -12.6$ at maximum. According to his calculations, when the apparent visual magnitude was 4, 2, and 0, the star would emit respectively 80,000, 500,000, and 3 million times more energy than the sun. Adopting for the emission a mean value of 500,000 times that of the sun, the total energy radiated in those 250 years turns out to be 1.5 × 10^{49} erg. This number tells us that during the period it was near maximum Eta Carinae must have emitted an amount of energy nearly equal

2. See *Beyond the Moon*, appendix A.

to that emitted by the sun in its entire life, and even this result is certainly an underestimate.

It now appears harder and harder to fit Eta Carinae among the known objects. In the past it was considered a slow nova, such as Nova Pictoris (1925) and Nova Herculis (1934), or an atypical nova like RR Telescopii. In support of this view, it was pointed out that at particular stages of their decline slow novae show the same lines of [Fe II] that appear in the spectrum of Eta Carinae. In slow novae, however, these lines are a transient feature, whereas in Eta Carinae they persist throughout. Furthermore, the amount of energy emitted by Eta Carinae in 250 years is at least 100,000 times greater than that emitted in a nova outburst, and even at minimum it emits almost as much energy as most novae produce in the few days of the explosion. Thus it is out of the question that Eta Carinae should be classified as a nova.

A comparison with supernovae is no less questionable. As previously mentioned, type I and type II supernovae have absolute magnitudes at maximum of -18.6 and -16.5, respectively. Having attained an absolute magnitude of -13, Eta Carinae would appear too faint to be included in either of these groups. On the other hand, the energy it emitted in 250 years is from 10 to 100 times greater than the energy expended in the outburst of a type I supernova but from 100 to 1,000 times smaller than that of a type II supernova. Moreover, a supernova outburst has an extremely rapid evolution, and to the extent that the explosion either destroys or radically transforms the parent star, it cannot be repeated. Eta Carinae exhibits instead slow variations which, as we mentioned earlier, may even include a periodic component. Its current brightening trend could be an indication of periodicity, which needs only time to be confirmed.

Undeterred by the above considerations, Zwicky, the famous supernova expert, assigned Eta Carinae to this class of objects. In our opinion, however, Zwicky was forced into it by his own broad definition of supernovae, despite the fact that he could not include Eta Carinae in any of the four groups he had previously defined. This in fact necessitated the creation of a fifth group to accommodate both Eta Carinae and the supernova discovered in NGC 1058 in 1961. We shall take up this question again at the end of the chapter. In the meanwhile we have new wonders to uncover that will shed some more light on our mysterious object.

THE CORE

Up to now we have considered Eta Carinae in its entirety and as a stellar object, the way it appears through a telescope at low magnification. Because their distance is so very great compared to their size, all stars appear pointlike even when viewed with the most powerful telescopes. Eta Carinae, instead, when observed with large telescopes and special techniques, does not appear as a point source. We shall now try to observe it more closely and investigate its structure in as much detail as possible.

The first person to notice any peculiarity in this respect was R. T. A. Innes who on May 10, 1914, discovered that the star had a very close companion of about 10th magnitude. Soon after he discovered another one, and more were later detected by other astronomers interested in binary stars. Concurrently it was ascertained that all these starlike objects, as well as Eta Carinae itself, were embedded in a small nebula. At first it was thought that Eta Carinae might be a multiple star like the Trapezium in Orion, and as such it was subsequently observed and measured by several astronomers. In 1950 E. Gaviola drew a chart of the region based on the best available plates at high magnification which enabled him to define fairly well the general appearance of the nebula and the different intensities of the various regions. The result was a strange configuration that he called "Homunculus," or manikin, as shown in figure 4.5.

In addition, Gaviola showed that the objects believed to be components of a multiple star system were actually condensations of the nebula in which the star was embedded. After gathering and comparing all the measurements performed on the star by observers of binary systems in over half a century, he concluded that all these condensations appeared to be moving radially away from the central star. From their apparent velocities (about 5 arc seconds per century), it could be determined that they had started to separate at the beginning of the nineteenth century.

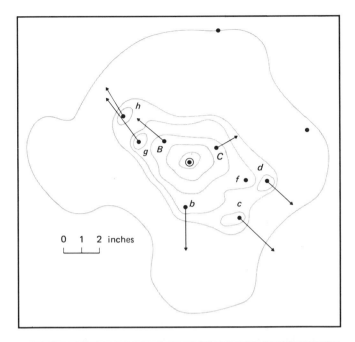

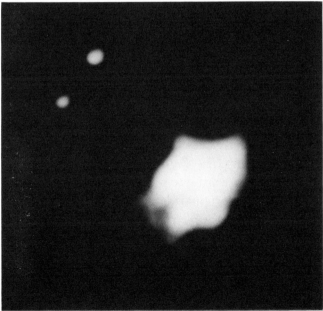

Figure 4.5 *Top*, an iso-intensity contour map of the Homunculus and the nebula around Eta Carinae (north is at the top). Brightness decreases from the center outward. The letters mark the stellar condensations that are moving, and the directions are indicated by the arrows. (Adapted from *Zeitschrift für Astrophysik.*) *Bottom*, the Homunculus photographed by R. D. Gehrz and E. P. Ney with the 1.5-m telescope at the Cerro Tololo Inter-American Observatory (north at top). (Courtesy of E. P. Ney.)

In addition to revealing a different facet of the star's behavior, this discovery afforded an excellent basis for computing its distance. First we derive the expansion velocity of the material ejected from the nucleus from the measurements of radial velocities obtained from the spectra. Then, by measuring the displacement of one or more condensations, we can find the increase in the apparent angular size of the nebula which is expanding at that velocity. Thus we obtain the number of kilometers corresponding to a given angle, or, in other words, the angle subtended from earth by a given length at the distance of Eta Carinae. From these two values we can readily derive the distance. For instance, if we assign a velocity of 590 km/s to region h (which in 1972 was 6.5 arc seconds from the center), we can compute that Eta Carinae is at a distance of 7,500 light-years.

On February 29, 1972, R. D. Gehrz and E. P. Ney took some photographs of the region, one of which we have reproduced in figure 4.5. The measurements they performed showed that the condensations identified by Gaviola were still expanding, and that in view of their displacement between 1944 and 1972 they must have originated in 1835, ±10 years. However, the expansion values found for the different condensations appeared to show a spread about a mean value of 1.25, suggesting that the material might have been ejected in successive phases over a long period of time. For instance, region h (the head of the Homunculus) would have left the central region around 1862. A subsequent reassessment of Gaviola's original plates by the same astronomers led to a slightly different date for the beginning of the expansion: 1849, ±7 years. Both results clearly indicate that the time of ejection of the nebula and of the starlike condensations embedded in it corresponds to the period of Eta Carinae's maximum brightness in 1843.

At the same time a new technique was bringing forth even more astonishing results. For several decades now the sky has been systematically observed also in infrared radiation. The first observations were carried out using photographic emulsions sensitive to the infrared. In 1965 the sky survey was extended to 2.2 microns and for several objects to even longer wavelengths. Systematic observations in the far infrared, at about 20 microns, started in 1968. In 1969 it was discovered that Eta

Carinae is the brightest object in the sky at this wavelength (outside the solar system). If our eyes had maximum sensitivity around 20 microns, rather than in the spectral range corresponding to yellow, we would see an entirely different sky. Many of the brightest stars, like Sirius and Vega, would become invisible, whereas other objects would stand out dazzlingly against the dark vault of the sky, notably the Omega nebula in Sagittarius, the center of the great nebula in Orion, the source IRC + 10216, the variable star VY Canis Majoris, the galactic center, the NML object in Cygnus, and many other infrared sources more or less intense, both starlike and extended. Still, Eta Carinae would outshine them all. Although only slightly more luminous than the Omega nebula, it would appear much brighter not only because of its greater total flux but above all because it is concentrated in a region nearly 7,000 times smaller.

Obviously then, we do not see this star in the most favorable conditions, as in the case of the sun. The distribution of the energy emitted by Eta Carinae is shown in figure 4.6; the broken line is an extrapolation based on the measured values at shorter wavelength and the assumption of a blackbody spectrum. On the basis of these observations, J. A. Westphal and G. Neugebauer concluded in 1969 that the absolute bolometric magnitude of Eta Carinae is now -12.0. As previously mentioned, Gratton found a value at maximum of $M_{bol} = -12.6$, which is only slightly higher. Are we then to conclude that Eta Carinae is radiating as much energy now as in 1843 when it was bright enough to rival Sirius? Westphal and Neugebauer think so and ended their 1969 paper with the following words: "One is tempted to suggest that this agreement [of the two values of bolometric magnitude], although perhaps fortuitous, is an indication that the absolute luminosity of Eta Carinae has remained essentially constant since 1843 and that the energy has been redistributed in wavelength since that time." Perhaps we cannot entirely agree with their conclusion, particularly because we do not know what bolometric correction should have been applied at the time of maximum in 1843. On the basis of the first spectrogram of the star (obtained in 1889) which showed the characteristics of spectral class F5, Gratton assumed no bolometric correction. However, we do not really know what spectral type the star was in 1843; moreover, as we have seen, its spectrum is very peculiar and does not remain constant. In a general way, however, it is reasonable to conclude that the energy emitted by Eta Carinae over its entire spectrum has not changed very much.

This point was better clarified by additional work reported by Gehrz, Ney, Becklin, and Neugebauer in 1973. Observing the star in the infrared at the various wavelengths possible between 2.2 and 18 microns, they found that the nucleus of the Homunculus itself not only is not starlike but exhibits a different diameter depending on the wavelength at which it is observed. At 2.2 microns its diameter is only 2 arc seconds, but choosing longer wavelengths up to 18 microns, the diameter increases up to 6 arc seconds. Given the diameter in arc seconds, and knowing the distance, we can then derive the diameter in kilometers; and from the wavelength at which the surface radiates, we can calculate the temperature. Thus we have learned that what we see in infrared is in fact a large dust shell 8,000 AU in diameter, or one hundred times larger than the solar system. The outer portion of this shell is relatively cold at a temperature of 250°K (corresponding to −23°C). The inner layers are progressively hotter. The innermost, observed at a wavelength of 2.2 microns, has a diameter of 2,600 AU and a temperature of 425°K, higher than scalding steam. Obviously the outermost envelope is also the coldest, since it is the farthest from the central body. It is this last envelope that gives the greatest contribution to the observed bolometric magnitude. Energy is supplied to it by radiation of a shorter wavelength emitted by a central object, which must be brighter in the visible but must also be of comparable bolometric magnitude.

Our overall view of Eta Carinae has now become much clearer. There is a nucleus, perhaps stellar, which emits a tremendous amount of energy, and from which gas streams out to form a nebular envelope or starlike condensations. Around this nucleus, perhaps because of the condensation of the ejected material, there are various dust shells that absorb the energy coming from the inner regions, starting from the core, and re-emit it at longer wavelengths. This model fails to explain the increase in brightness observed in the last century. At this point we might speculate that perhaps it never really took place and that the radiation has always been more or less the same. When the dust envelope

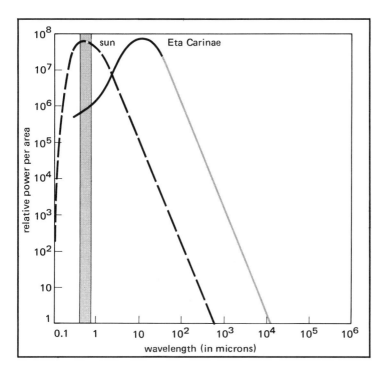

Figure 4.6   Energy distribution of radiation from the sun and from Eta Carinae. The shaded strip represents the region of wavelengths in the visible, from ultraviolet to far infrared. The gray portion of Eta Carinae's curve is a guess: no measurements exist. (From G. Neugebauer, E. E. Becklin, *Scientific American*, April 1973.)

forms or thickens, the radiation is mostly in infrared; when the envelope dissolves, possibly as a result of processes occurring in the nucleus, the radiation comes directly from the nucleus itself in the form of light.

NEW VISION

THE LOCATION OF ETA CARINAE

Up to now we have described the appearance of Eta Carinae and attempted to understand its nature and structure. Following this approach, we have made good progress, but there still remains the fundamental problem of determining what such an object could be. We have seen that it was regarded first as a nova and then as a supernova and that there are valid arguments against both hypotheses. We will now attack the problem more directly, using the knowledge we have acquired and introducing an important factor that has not been considered so far—the study of the region where Eta Carinae is situated.

As we said at the very beginning, Eta Carinae is in a region of the sky rich in blue stars where there is also a beautiful extended nebula known as NGC 3372. What we do not know yet is whether the blue stars, the nebula, and Eta Carinae are all at the same distance from us, that is, whether they are associated with one another or simply appear to be in the same region because of projection effects. A number of astronomers have tackled this question at various times. L. Gratton, in particular, carried out around 1960 an in-depth study of the region and reached an astonishing conclusion that has been confirmed by the most recent observations.

In 1949, when almost all the O associations were defined and classified, it was found that most of the blue stars near Eta Carinae form a rich and extended association which was given the name of CAR OB 1. B. J. Bok had already shown in 1932 that most of the blue stars belonging to this association are embedded in the nebula NGC 3372 and are responsible for its luminosity. Consequently, it was quite important to find out whether Eta Carinae was also embedded in the nebula and belonged to the CAR OB 1 association or was instead a background or foreground star. Gratton solved the problem, bringing to an end the long controversy over the interpretation of J. Herschel's observations.

During his stay at Cape Town Herschel observed Eta Carinae and mapped the entire region with great accuracy, drawing a picture of the nebula and the stars as he saw them with his telescope. His illustration (figure 4.7a) clearly shows that at the center of the nebula there was a sharply defined dark area in the shape of a keyhole. This feature was so characteristic that the entire nebula became known as the Keyhole. Thirty years later the appearance of the nebula changed; one side of the keyhole appeared open, as can be seen in present-day photographs (figure 4.7b). This finding generated a heated controversy. Herschel, who was still living at the time, stated that he had drawn the region exactly as he had seen it and had defined its contours with great precision with the help of some stars located at the edges of the dark nebulosity, just as shown in his drawing. Today these stars define only the upper part of the Keyhole (see figure 4.7b), while the lower half of the border is missing because a large portion of the bright nebula drawn by Herschel has totally disappeared. Gratton demonstrated that Herschel had not merely imagined this part of the nebula. In fact, it could be photographed using modern equipment and long enough exposures capable of revealing very faint features below the limit of sensitivity of Herschel's telescope. The part of the nebula that Herschel saw and drew really exists, but at the time Herschel drew it it must have been much brighter than it is today. According to Gratton this fact can be explained in a very simple way. Since this region corresponds to the environs of Eta Carinae, it was Eta Carinae—then at its brightest—that either lit up or excited the nebula. This proves that Eta Carinae is indeed embedded in the nebula. Since the nebula is part of the association, as demonstrated by Bok, it follows that Eta Carinae must belong to the association as well.

This last point was also confirmed by Gratton and his collaborators. By obtaining the spectra of about fifteen stars that are undoubtedly part of the association, they were able to show that the H and K lines of interstellar calcium appearing in them have the same intensity as those found in the spectrum of Eta Carinae. Hence Eta Carinae must be at the same distance as the stars of the association.

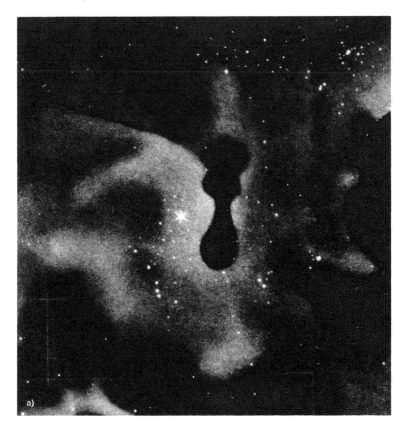

Figure 4.7   *Left*, drawing of the region around Eta Carinae by J. Herschel. *Right*, actual photograph obtained with the 1.52-m telescope at Bosque Alegre, Argentina. The white disks mark the stars Herschel found to be at the border between the bright and dark regions. Note the different appearance of the nebula a century later. This may be due to the varying intensity of Eta Carinae which illuminates and excites the interstellar gas. (Courtesy of the Specola Vaticana and L. Gratton.)

This result was further refined in 1975 by N. R. Walborn and J. E. Hesser. From the high-dispersion spectra of sixteen stars (mostly class O), which like Eta Carinae appear superposed on the nebula NGC 3372, they discovered that the profile of the K line of calcium is simple and regular in stars far from the central region but increasingly more complex in the spectra of stars nearer the center, where Eta Carinae is also situated. The complex structure of this line, formed by interstellar calcium, is due to the fact that the calcium clouds between us and the star are moving at different speeds. On the other hand, the lines appear to be regular in stars at the periphery of the nebula. Thus at least part of the calcium in which the observed absorption takes place belongs to the nebula itself, and not to the intervening space as was believed at first.

Eta Carinae, which shows as complex a K line as that of the other stars in the nebula, must therefore be part of the group or, at worst, be located behind the nebula. In this case, however, it would have to be even farther away, and therefore even more luminous, which further complicates its interpretation. Moreover, the most recent observations agree in placing Eta Carinae in the CAR OB 1 association.

As we know, associations consist of very young stars, either just born or beginning to form from nebulous matter present in the region. Gratton concluded that Eta Carinae might well be a very young star—perhaps one at the very beginning of its evolution. The initial phases of stellar evolution have never been observed in any other star and are still far from clear. In effect the monstrosity of Eta Carinae may be only that of embryonic matter which has yet to assume its final form. What we unfairly call a monster of the sky might one day produce a star.

WHAT IS ETA CARINAE?

Gratton's conclusions, of which we have presented only the bare essentials, seem to have put us on the right track. We will continue now with the aid of the latest findings. The most significant of these is the increasingly clear evidence of intense activity from the entire region. In 1970 radio astronomers discovered the existence of a gaseous ring expanding at the speed of 20 km/s. This ring corresponds roughly to the upper part of the Keyhole. If the velocity had always been the same, the expansion would have started approximately 10,000 years ago. Last year, moreover, N. R. Walborn published the results of a careful optical study of this same region. He showed that the radio ring is associated with an intense emission arc due to [S II]. Figures 4.8 and 4.9 show the region at different magnifications. Figure 4.9 also shows the center of the expanding ring, as determined by the radio observations. Monochromatic plates of great interest were obtained by L. Deharveng and M. Mauscherat, who published them in the second half of 1975 together with measurements of radial velocities. These photographs clearly show strange configurations which are typical of diffuse matter in the earliest stages of evolution, such as the so-called "elephant trunks" and Bok's dark globules. Thus there is no longer any doubt that this region, like that observed in the Orion nebula, is a veritable forge of stars.[3]

As it happens, in the last few years strange infrared objects have been discovered in Orion, called by English-speaking astronomers "cocoon stars." These objects emit a tremendous amount of energy, but only in the infrared. Each of them probably corresponds to a very hot star that is forming and is still surrounded by a dust envelope dense enough to absorb nearly all of the visible and ultraviolet radiation emitted by the star. Some astronomers believe that the absorption caused by clouds of this kind may be as large as 25 magnitudes.

To understand the enormity of this absorption consider that if the light of the sun were to be reduced by the same amount, the sun would appear only slightly brighter than Sirius. With such a dust cloud surrounding the sun, we would see the stars night and day. The night sky would actually remain the same, but at certain times there would rise on the horizon a dark, nearly circular patch with a disk at the center as large as the full moon, yet very faint. Under these conditions the moon would be completely invisible, and obviously life as we know it would not exist on earth.

The energy the cloud absorbs from the star raises its temperature to about 500°K, which corresponds to an intense infrared emission. Eta Carinae could also be surrounded by a similar cocoon—but one less than

3. See *Beyond the Moon*, chapter 5.

Figure 4.8   The region surrounding Eta Carinae photographed with the 4-m telescope of the Inter-American Observatory at Cerro Tololo, Chile. In this plate several bright stars, an open cluster (top right-hand corner), and both bright and dark nebulae can be seen. (Courtesy of N. R. Walborn.)

Figure 4.9 The photograph in figure 4.8 is reproduced here at higher magnification. The small black circle marks the center of an expanding gas ring discovered in 1970 by radio astronomers. Eta Carinae lies on the line going from the black circle to the lower left corner of the photograph and is far from stellar in its shape. (Courtesy of N. R. Walborn.)

perfect, perhaps in a more advanced phase of evolution. At certain times or in certain places, tempestuous activity in the central body may cause this cocoon to open up or dissolve, affording us a glimpse of the inner regions, that is, the crucible where the star is forming. Using all the information gathered to date, we can now attempt to determine what type of star is forming in the cocoon. This brings us to the last surprise in the story of Eta Carinae.

We recall that Eta Carinae has a $M_{bol} = -12$, and is surrounded by a dust cloud 8,000 AU in diameter. This means that the object within the cocoon emits as much light as 3,400,000 sunlike stars. Of course, not all of these stars actually need to exist. If they did, they would be clustered in an extremely small volume. The number of stars in the space surrounding the solar system is much smaller, only 4 stars to a volume 32 million times larger than that of Eta Carinae's cocoon. An exceptionally high concentration of young stars in the Eta Carinae's region would be one more proof that stars are indeed being born there. Where does one find the highest number of newborns? In the maternity ward, of course.

As we said, however, it is not at all clear that over 3 million sunlike stars are being born in Eta Carinae. They could all be much brighter stars, and therefore far less numerous. For instance, if they were all B0 stars, their number would decrease to little more than 100, while in the case of all O stars, the brightest we know, 10 would be quite sufficient.

What seems fairly evident at this point is that something is being born in Eta Carinae, and this something is almost certainly not a single star. It may be the kernel in which a multiple system of young stars is forming, such as those we find at the center of the Orion nebula (the so-called Trapezium) and in the Trifid nebula. These systems are more extended, however, possibly because they are in a more advanced phase of evolution. It is even possible that the cocoon may contain a whole group of 100 to 200 stars, or more, provided that among the brightest stars there are some less massive ones. One day the dust will be blown away by the radiation emitted by the stars, and the cocoon will dissolve. We will then behold a brand new group of stars that will gradually pull away from one another to take separate paths in the vastness of space.

IS ETA CARINAE REALLY UNIQUE?

We said earlier that Eta Carinae is the only object of this kind that we have observed in the entire Galaxy. Does this mean that it is truly unique in the whole universe? Probably not. We have not explored the entire Galaxy, and it is possible that similar objects exist in regions that are inaccessible to us, either because of their great distance or because interstellar clouds obscure them from us. In addition, other objects of this type might be in different evolutionary stages or have different structures. Hence they would appear similar to Eta Carinae but not exactly the same. If Eta Carinae is a star, or group of stars in the process of formation, as Gratton concluded, it certainly cannot be unique. On the other hand, it is just as certain that if not unique, Eta Carinae is clearly the prototype of an extremely rare class of objects.

We will first look for similar objects in our own Galaxy. Here, however, poor knowledge of distances and possible occultation by interstellar clouds might cause some problems. We will then extend our search to external galaxies where even at great distances discovery is still possible because of the very high luminosity of these objects.

SLOW NOVAE

Students of variable stars specializing in novae have defined a small subclass—the "slow novae"—which is characterized mainly by the fact that the phenomena observed to occur in a few days in the light curves and spectra of ordinary novae are protracted instead over months or years. This group, to which at a certain point Eta Carinae had also been assigned, consists of a very few stars. We will examine only three of them—RT Serpentis, RR Telescopii, and P Cygni—mainly from the standpoint of possible similarities with Eta Carinae.

RT Serpentis. This star is the prototype of the group, and of the three the most like a true nova. It was observed for the first time in 1919, when it was already at maximum, but its light curve could be reconstructed fairly well from plates of the field that had been previously taken for other reasons. It was found that prior to 1909 the star was invisible and therefore must have been fainter than the 16th magnitude. On July 9, 1909, it was recorded for the first time on a photographic plate, at magnitude 13.9. In 1910 it brightened to the 11th magnitude. The following year it faded slightly by 0.6 magnitudes, then started slowly to brighten again, attaining maximum in 1914. For the next nine years it was practically constant between magnitudes 10.3 and 10.6. In 1924 it started to get dimmer and by 1941 had faded to magnitude 13.5 (figure 4.10).

RT Ser's spectrum exhibited the normal development observed in novae, although it had far longer duration. In 1919, when the first spectrograms were obtained, it showed absorption lines that appeared weak, owing to the superposition of an emission component already increasing in intensity. During the remaining period of maximum (through 1923) there developed the Balmer series of emission lines. In 1928 the forbidden lines of iron appeared in emission, which are typical of slow novae as well as of Eta Carinae. Finally, three years later, the so-called nebular lines could be observed, a further characteristic of a specific phase of all novae. Spectroscopic observations carried out up to 1942 revealed the presence of lines corresponding to ever increasing excitation. A similar behavior will be later observed in RR Telescopii.

In view of its spectral characteristics, RT Ser could be considered a nova. Nevertheless, it is hard to explain how phenomena that in normal novae are due to explosive events could appear in this star from a slow ejection of gas which retraced over many years the same phases that normally take place in a few months.

RR Telescopii. Even more interesting and better studied is the case of RR Tel, discovered in 1908 at the Harvard Observatory from a score of plates obtained between 1894 and 1907. At first the star did not appear as an exploding nova but rather as a periodic variable, fluctuating between magnitudes 12 and 17 in slightly more than a year. Subsequently, on the basis of a greater number of observations, S. Gaposchkin classified the star as a semi-regular variable with a period of 387 days.

In October 1948 the star appeared exceptionally bright, that is, of the 7th magnitude. An examination of its light curve (figure 4.11) shows that the increase in brightness, from m 14 to m 7, began in October 1944

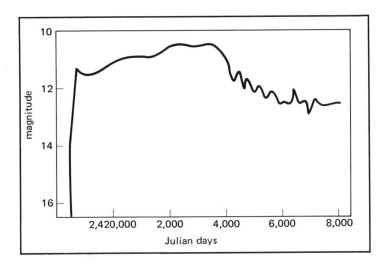

Figure 4.10   Light curve of the variable star RT Serpentis from 1908 to 1935. (From C. Hoffmeister, *Veränderliche Sterne.*)

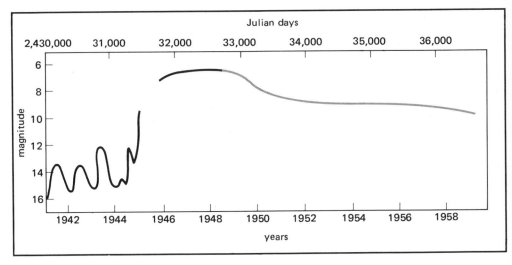

Figure 4.11   Light curve of the variable RR Telescopii from 1941 to 1958. The photographic observations are in black, and the visual observations in gray. (From *L'Astronomie.*)

and that the star remained at maximum for over 3 years. From mid-1949 it started to fade, and by 1957 it had decreased to the 9th magnitude.

Spectroscopic observations started on May 6, 1949. Of particular interest are the observations carried out by A. D. Thackeray, director of the Radcliffe Observatory near Pretoria, South Africa. For more than 10 years he obtained spectra of both RR Tel and Eta Carinae with the 189-cm telescope at the observatory. In 1953 he published a comparative study of the two stars. This revealed that from May to June 1949, while at maximum, RR Tel had an absorption spectrum of cF5 type, similar to that of Eta Carinae in 1893 and to the spectra of some slow novae at maximum. From June to October 1949, when the light curve had begun to decline, there appeared emission lines that gradually took the place of most of the absorption lines. Between 1951 and 1952 nebular lines appeared and strengthened, and the development of the entire spectrum was indicative of increasing ionization of the atmosphere of the expanding star, which also continued later on. Using the absorption lines of helium on the violet edges of the emission lines (P Cygni effect), Thackeray computed the ejection velocity of the gas, which turned out to be 685 km/s in 1951 and 865 km/s in 1952, of the same order as the velocities of nova envelopes.

Of particular interest was a comparative study of the time scales for evolution of RR Tel, RT Ser, and Eta Carinae based on some specific reference epochs, such as the time of the first increase in brightness, the time of maximum, the time of appearance of certain lines, and others. Using as reference the evolution of RR Pictoris, fastest of the so-called slow novae, Thackeray found that RR Tel had been 6 times slower, RT Ser 17 times, and Eta Carinae 220 times slower. RR Tel, in particular, turned out to be 100 times slower than a typical rapid nova like CP Lacertae, which reached maximum in 1936. In other words, it took RR Tel 27 years to complete the development undergone by CP Lac in 100 days.

The most significant similarity that Thackeray found between RR Tel and Eta Carinae was the appearance in both stars of forbidden lines, notably those of [N II], [S II], and [Fe II]. Finally, an interesting comparison was made by Gehrz and his collaborators in the 1973 work previously mentioned. Having discovered that Eta Carinae is the brightest infrared star in the sky, they observed RR Tel with the same instrumentation to determine whether it would exhibit intense infrared emission as well. The result was negative. Thus RR Tel does not appear to be surrounded by a dust envelope, and at least from this standpoint is quite different from Eta Carinae. No similarity could be found in the series of spectra obtained at the Argentine Observatory at Bosque Alegre during a period, around maximum, not covered by any other observations. The results of this research, which were never published, further excluded any similarity between Eta Carinae and RR Tel, since the behavior of RR Tel at the time was identical to that of novae.

P Cygni. This star was never mentioned by ancient astronomers, and it is believed that they never observed it. In 1600, however, it became easily visible to the unaided eye, although it did not attain the intensity that Eta Carinae was to exhibit two centuries later. The first to notice it was the Dutch astronomer W. Blaeuw who about 1640 marked the star on a celestial globe and wrote: "The new star in Cygnus that I was the first to observe on August 8, 1600,[4] was initially of the 3rd magnitude. I determined its position . . . by measuring its distance from Vega and Albireus. Its position has not changed, but the star is now no brighter than the 5th magnitude." After discovery, the star remained quite bright for six years; then it faded slowly to the 6th magnitude, recorded in 1620, and by 1626 was no longer observable with the naked eye. It brightened again around 1655 and hovered around magnitude 3.5 until 1659. Three years later it vanished once more but brightened somewhat in 1665. After undergoing several more variations in brightness, in 1715 it stabilized at magnitude 5, and has remained stable ever since, except for small fluctuations not in excess of 0.3 magnitudes.

P Cygni is of spectral type B1eq. The notation eq indicates the star exhibits an anomaly that is sometimes observed also in the spectra of other

---

4. This date is expressed according to the Julian calendar because at that time the Low Countries had not yet adopted the Gregorian calendar. It corresponds to August 18.

stars, some of which we have already mentioned, and in that of Eta
Carinae itself. It is the so-called P Cygni effect. Several lines, including
the Balmer series of hydrogen, appear in emission and broadened. Next
to the emission line, shifted to the violet, there appears a narrow absorption line corresponding to the same element. This phenomenon, easily
visible in the spectrum in figure 4.12, can be explained in a way that is
both simple and intriguing.

Suppose we have a star of rather high temperature, for instance one of
spectral type B, surrounded by a nearly transparent envelope, which in
the most common case will be hydrogen. By the mechanism we have described for planetary nebulae, the star excites the hydrogen atoms which
will then emit their characteristic lines, for example, H_α, H_β, H_γ. The
presence of the envelope is thus revealed by the lines of the Balmer se-
ries, which will appear in emission and narrow. Let us assume now that
the envelope is expanding (figure 4.13) and that all its points are moving
at the same velocity. The observer on earth who measures only the radial
velocity, the component of the expansion velocity in the line of sight, will
see the various parts of the envelope moving at different speeds. To him
the B points will appear stationary; the C points will be approaching at
a velocity that is only a fraction of the expansion rate, while the D points
will move away at the same speed; finally, A will appear to approach
at exactly the expansion velocity. The region behind the star comprised
between the two thin lines is hidden from view and therefore does not
make any contribution. Thus a hydrogen line that would appear narrow
for a stationary envelope will now be divided into many components,
some of which are shifted to the red and others to the violet depending
on whether they originate from points that are moving away or coming
toward us. The result will be a broadened emission line whose center
corresponds to the regions of the envelope that do not appear to move.
The greatest velocities both away from and toward us, and therefore the
largest shifts to the red or the violet correspond to the regions around A'
and A. The former region, as we said, cannot be seen, and the latter will
produce absorption lines since it is located in front of the star.

The appearance of broadened emission lines accompanied by narrow
absorption lines of the same element shifted to the violet is therefore in-

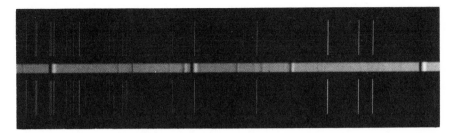

Figure 4.12   Spectrum of P Cygni obtained by G. Natali and R. Viotti with the 193-cm telescope at the Haute-Provence Observatory. As usual, reference laboratory spectra are given above and below.

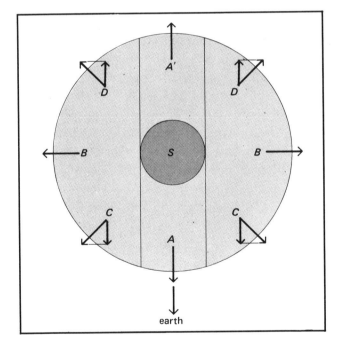

Figure 4.13   The P Cygni effect produced by an expanding envelope surrounding a star, S. In the region between the two straight lines and facing the earth, the gas absorbs the radiation emitted by the star. This produces a narrow dark line shifted to the violet by an amount that depends on the expansion velocity of the envelope. The regions to the right and left produce an emission line shifted to the red of the absorption line and broadened by a red shift that differs from point to point.

dicative of an expanding gas envelope consisting of that very element. Furthermore, by measuring the shift of the absorption line with respect to the center of the emission line, one can compute the expansion velocity of the envelope in kilometers per second.

In P Cygni this velocity ranges from 50 to 240 km/s, depending on the distance of the expanding atmosphere. It increases in the outer layers of the shell which expand at a greater rate than the inner ones.

P Cygni is very far away. It has been ascertained that this star is located in one of the spiral arms of the Galaxy, that in Cygnus. Consequently, it should be at a distance of at least 7,000 light-years; allowing for absorption, its absolute magnitude appears to be −8.9. Although P Cygni is in a seemingly quiescent stage, its brightness at the present time equals that attained by a nova in the few days of maximum light. When it was at maximum, it must have attained magnitude −11.9, nearly the same as Eta Carinae's. It is only because P Cygni is farther away from us that its brightness in our sky could not rival that of Eta Carinae.

We have now finished our examination of the three galactic objects that are most similar to Eta Carinae. We have described and discussed their characteristics not so much to find out to what extent Eta Carinae might resemble a nova—a rather unlikely possibility after all we have learned—but to point out a number of stars currently classified as novae (albeit special!) which might in fact belong to a separate class of objects together with Eta Carinae.

It is also possible that the objects in this special group, although exhibiting similar phenomena, will turn out to be intrinsically different both from novae and from Eta Carinae itself. For instance, while it is true that the P Cygni effect reveals an expanding envelope, it is not clear that such expansion is always caused by the same process. In one case, it may result from a nova explosion due, as we currently believe, to its membership in a binary system. In another case, it may be caused by a supernova explosion, that is, by the self-destruction of a star that has reached the end point of its evolution. Finally, the ejection of the outer layers may correspond to a process that takes place in a star just being born. The P Cygni effect is only one of the observable features. There may be others, such as the light curve, the infrared emission, and the characteristic line spectrum, all of which may appear similar but are due to different causes.

To draw general conclusions from preliminary and often incomplete results on the basis of only a few well-known objects can be dangerous and may lead to totally erroneous deductions. This happened, for instance, at the beginning of the century when the Cepheids were interpreted as elipsing binaries, whereas they are actually pulsating stars.

It is already quite interesting that we have found these four exceptional objects. We may be able to add to their number and to understand them further by looking for stars similar to Eta Carinae in external galaxies.

TYPE V SUPERNOVAE

On July 11, 1961, in the course of a systematic survey carried out at the Bern observatory, P. Wild discovered a 13.5 magnitude supernova in the galaxy NGC 1058. When news of this discovery reached Asiago, F. Bertola promptly undertook spectroscopic and photometric observations of the star. Since they were the only ones in the world performed at the time with adequate instruments, these measurements turned out to be considerably valuable. It was immediately apparent to Bertola that the spectrum was quite different from those of both type I and type II supernovae. But an even more peculiar fact came to light. From earlier plates of the region it was found that the star was already visible in 1937 and that it had remained at about magnitude 18.0 until 1954. Only one plate could be found for the period December 21, 1954, to July 11, 1961 (obtained on November 21, 1960), and it showed the star at magnitude 15.8. Evidently, the star had already begun to brighten, but we cannot define the exact duration of the increase in luminosity that brought it to a brightness sixty times greater than it had been for at least twenty years.

Such behavior had never been observed before. No supernova had ever exhibited a period of quiescence prior to its rise to maximum, which is always extremely fast. The surprises were by no means finished because the star had not yet attained true maximum. Following Wild's discovery, it remained stationary for over three months, and even faded a little, then brightened rapidly to magnitude 12.2, an increase of 1.3

magnitudes. Henceforth it started to decline, but its light curve was not in the least regular, as shown in figures 4.14a and b. It was photographed for the last time on March 6, 1970, by F. Bertola and H. Arp with the 5-m telescope of Mount Palomar. At that time the star had faded to magnitude 21.7, and was therefore 6,300 times fainter than at maximum and 30 times fainter than it had been during the long pause at around magnitude 18 (1937 to 1954) before the final upward push.

On the strength of several arguments it was shown that this star belongs to NGC 1058 and does not just appear to be there because of random superposition. Assuming a distance module for that galaxy $m - M = 29.0$, and allowing for galactic absorption, Bertola found that the star attained at maximum $M_{pg} = -17.2$, an intermediate brightness between a type I and a type II supernova. Its spectra, on the other hand, were in no way similar to those of either supernova type. They showed the lines of hydrogen, He I, Fe II, and C III in emission, with a broadening that indicated the envelope was expanding at a speed of 3,700 km/s. In certain periods absorption lines were also observed, but the forbidden lines did not appear at all.

The observations carried out by Bertola and Arp between 1968 and 1970, when the star was already fainter than the 21st magnitude, disclosed a very significant fact. In the four plates obtained with the 5-m telescope the star appeared embedded in a small nebulosity (figure 4.15). This nebulosity could not have been detected earlier because of the greater luminosity of the star, and there may be at least two explanations for it.

One possibility is that it is due to material expelled from the star during the explosion. Since no forbidden lines had appeared in the spectrum, Zwicky had already predicted that the mass and density of the gas would be much greater than those of the envelopes of ordinary novae. On the basis of the known distance ($m - M = 29.0$) and expansion velocity ($V_{exp} = 3,700$ km/s), the gas cloud could have attained a large enough diameter for us to be able to detect it (at least 1 arc second) only if it had been expelled 10,000 years ago. Thus, if the nebulosity had formed from a supernova explosion, it would have had to last a very long time or be recurrent. We know of no supernovae with either characteristic.

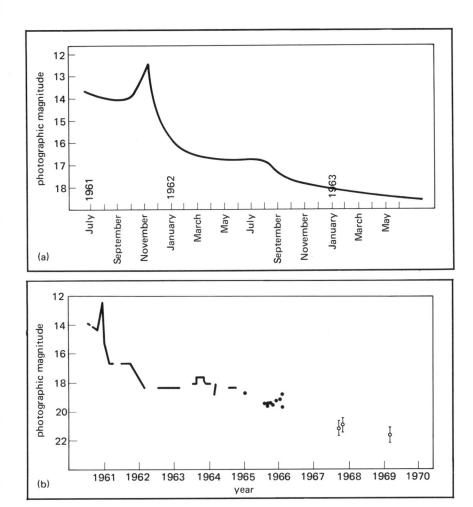

Figure 4.14 (a) Light curve of the supernova in NGC 1058 from July 1961 to July 1963. (b) Light curve of the same supernova over a longer time span, from 1961 to 1970. (From *Colloque Internationale sur les Novae* and *Publications of the Astronomical Society of the Pacific.*)

Another way to explain the nebulosity is to assume that it always existed and that the bright star appeared at the very heart of it. As we have observed in both our own and external galaxies, as well as in the case of Eta Carinae, groups of young stars are often associated with emission nebulae and interstellar matter. These systems are found mainly in spiral arms where they appear as knots or condensations. Now, by combining the four plates obtained with the world's largest telescope, Bertola and Arp obtained a single image which showed that both the star and the nebular condensation were at the end of a thin filament protruding from the galaxy like a spiral arm, along which other condensations were also visible.

It is possible therefore that the star may belong to an association and that it may be a very young object, such as Eta Carinae appears to be. There is a certain similarity in the respective light curves, even though the object in NGC 1058 evolved much more rapidly, which is quite consistent with the larger expansion velocity of the envelope (3,700 km/s instead of less than 600). The two objects differ substantially, however, with respect to their absolute magnitude at maximum and, more importantly, in the question of the forbidden lines. Whereas the forbidden lines are one of the main features of Eta Carinae's spectrum, they do not appear at all in the spectra of the star in NGC 1058.

This star has been commonly regarded as a supernova, particularly by Zwicky. However, being well aware of the difference between its characteristics and those of other types of supernovae, he assigned it together with Eta Carinae to a special group, type V. Previously he had assigned to this class, first introduced in 1963, three other supernovae: that of 1909 in M 101, and those of 1923 and 1957 in NGC 5236. They are not listed as type V supernovae in the latest catalogue, but described as "peculiar," and only two other supernovae are indicated as probably belonging to type V: that of 1954 in NGC 2403 and that of 1964 in NGC 3631. Consequently, Eta Carinae and the 1961 object in NGC 1058 remain as the only certain members of type V. In Zwicky's view they are both supernovae, perhaps of a very rare type.

This conclusion is somewhat surprising. If you recall, supernovae are stars at the end of their evolution which self-destruct in a sudden and very rapid explosion. When this occurs, they are transformed into completely different objects, namely, pulsars or neutron stars, and may produce very conspicuous envelopes. We should also recall that they are so faint in visible light that normally, even in the case of galactic supernovae, we cannot see the objects themselves, although we can often detect radio and X-ray emission in the region. Neither Eta Carinae nor, as far as we know, the star in NGC 1058 appears to have these characteristics. One cannot help wondering, therefore, why Zwicky would regard them as supernovae.

The answer lies in his own definition of supernovae. As he stated in the 1963 Colloquium on Novae and Supernovae at the Observatory of Haute-Provence, it is his working assumption that a supernova is a star that (1) attains at maximum a visual luminosity M_r greater than -12 and (2) emits in the visible at the time of maximum a total energy larger than 10^{48} erg every ten years.

Supernovae of all the types defined by Zwicky, including those of type V and Eta Carinae itself, satisfy this definition. On the other hand, if by supernova we mean an exploding star that transforms itself radically in the course of a few days, we cannot apply this name to Eta Carinae since it almost certainly has been radiating the same amount of energy for at least two centuries. Naturally, Zwicky was aware of this difficulty and tried to cope with it by introducing types III, IV, and V. However, the new groups came to include some supernovae that might belong to type I or II, as well as objects that might not be supernovae at all. Consequently, Zwicky's new groups have not found much favor among astronomers. In particular, let us consider type V which is relevant to our present discussion. In our opinion, there may be variable stars that satisfy both conditions set by Zwicky, without being supernovae in the classical sense, and could constitute instead a different class of variables. Moreover, some variables that could belong to this new class might be excluded on the ground that they do not satisfy Zwicky's conditions. For instance, even if we considered type V as a new category of stars having nothing in common with supernovae, we could not include in it all those

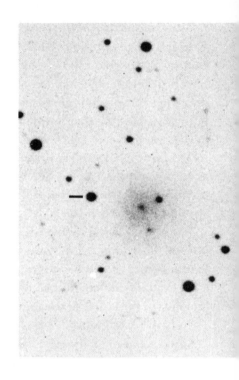

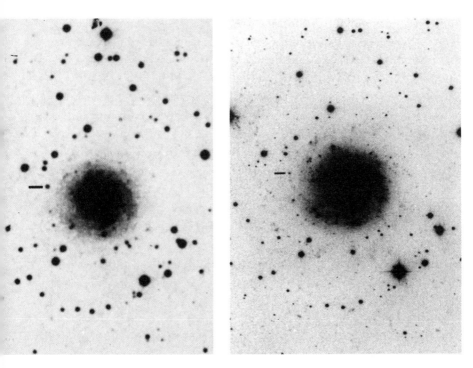

Figure 4.15   *Left*, NGC 1058 with the supernova at maximum photographed on December 7, 1961, with the 1.20-m telescope at Asiago. *Center*, the same supernova at intermediate magnitude photographed on September 8, 1962, with the same telescope and longer exposure time. *Right*, the supernova about to disappear in a photograph obtained on October 21, 1968, with the 5-m telescope at Mount Palomar and even longer exposure time. In this picture the spiral arm of the supernova can just be seen. (Courtesy of F. Bertola.)

objects that did not attain at maximum $M_v = -12$, although typical in every other respect. P Cygni, for example, is so similar to Eta Carinae that it should be part of the same new group of variables but must be excluded because at maximum it attained $M_v = -11.9$.

For all these reasons it is perhaps inappropriate to call supernovae objects that have only certain aspects in common with true supernovae. It is already unfortunate that the word "supernovae" should suggest a similarity with "novae," which are in effect very different objects. The difference between type I and type V supernovae is perhaps even greater. Consequently, rather than giving the same name to objects as far apart as the Crab nebula and Eta Carinae, it would seem more appropriate to segregate all stars of the latter type into a new class of high luminosity variables. Even though at the moment we may not be able to explain the few we know, this appears advisable in the expectation that our understanding will deepen in the future.

HUBBLE-SANDAGE VARIABLES

At the end of our discussion of type V supernovae, we mentioned the possibility of forming a new class of variable stars. In fact, such a class might already exist.

In the course of a study of five variables (one in the galaxy M 31 and four in M 33), E. Hubble and A. Sandage called attention in 1953 to a new type of variable stars characterized by high luminosity, blue color, and spectral type F. Their light curves (figure 4.16) exhibited an odd behavior: small fluctuations; a slow rise to maximum light, followed by a steep decline; or a period of quiescence followed by a more or less slow rise to maximum and subsequent decline. Common to all five stars was the very slow evolution of their light curves: they evolved over years, whereas the slowest normal variables, the so-called long-period variables, evolve over months. As a rule the amplitude of these oscillations did not exceed 3 magnitudes. The most startling feature of these stars was their brightness at maximum, which on the average was found to be $M_v(\text{max}) = -8.4$. No stable star of such high luminosity had been previously seen.

Subsequent work confirmed and emphasized the exceptional charac-

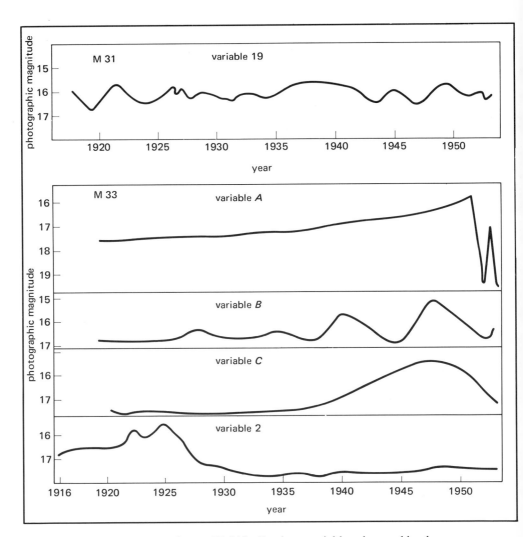

Figure 4.16 Light curves of some Hubble-Sandage variables observed by the two astronomers in the galaxies M 31 (top) and M 33. (Adapted from *The Astrophysical Journal.*)

ter of these stars. In 1973 L. Rosino and A. Bianchini announced the discovery of eight additional variables of this type—three in Andromeda (M 31) and five in Triangulum (M 33), both of which are among the nearest galaxies. One of these variables, Var A-1 in M 31, turned out to be particularly interesting. Its light curve showed the typical fluctuations and, in addition, a slow decline. Searching old photographic plates, the Russian astrophysicist A. S. Sharov was able to reconstruct its light curve from 1888, and he found that at that time the star was more luminous. In eighty-four years it had declined slowly from magnitude 14.8 to 16.3. On the basis of Andromeda's distance, in 1888 the star's absolute visual magnitude must have been $M_v = -10.3$. This gives us a lower limit to its absolute magnitude at maximum. However, as we pointed out, we are observing the object in a declining phase, and it is possible that we never saw its true maximum, which therefore may have reached an even higher value.

The story of this object does not end here. In February 1975 Rosino and Bianchini obtained a few spectra with the 1.82-m reflector of Asiago Observatory (Cima Ekar). To their amazement, they found that these spectra were remarkably similar to those of Eta Carinae, in both the continuum and the emission lines of H, He I, and Fe II. They were particularly struck by the similarity in intensity distribution of the continuum and by the large number of Fe II lines, both permitted and forbidden, which were already a well-known feature of Eta Carinae's spectrum.

Tamman and Sandage had previously found another star similar to Eta Carinae in NGC 2403. Its light curve (figure 4.17) exhibited irregular fluctuations from 1910 to about 1954; then it began to rise with wider and more frequent oscillations. In November of 1954 the star attained an absolute visual magnitude of -12.3 (or perhaps higher). From then on it kept fading, and in early 1963 it was no longer observable even with the world's most powerful telescope.

Thus at least two of the Hubble-Sandage variables have turned out to be similar to Eta Carinae, and we cannot exclude the possibility that the others might also be. While Rosino and Bianchini were carrying out their research at Asiago, R. M. Humphreys independently obtained the spectra of five of these variables with the 2.30-m telescope of Steward

Observatory and with the 2.10-m telescope of Kitt Peak. Three of the variables belonged to M 33 and two to M 31. One of the latter (AE And) exhibited an emission spectrum nearly identical to that of Eta Carinae (figure 4.18). The others had enough characteristics in common that, according to Humphreys, all five stars should be considered of the same type as Eta Carinae. None of these objects attained the exceptional maximum brightness of either Eta Carinae or the two variables observed by Bianchini-Rosino and Tamman-Sandage. Therefore they appear to be normal Hubble-Sandage variables with light curves entirely different from that exhibited by Eta Carinae in the last century.

According to Rosini and Bianchini, the current absolute photographic magnitude of these objects is, on the average, −8.8 for M 31 and −9.0 for M 33. It may be very significant that in the quiescent period prior to its rise to maximum, Eta Carinae's absolute magnitude was also −8.

From all we have learned in the past few years, Hubble-Sandage variables appear to be stars like Eta Carinae observed in a quiescent period. One is now struck by an interesting thought: should these variables, observed at about absolute magnitude −9, brighten by 4 magnitudes or more, they would fully satisfy Zwicky's definition of supernova. As we have seen, this is what actually happened in two different cases, but the objects were not called supernovae. On the other hand, one would have been forced to classify as supernovae the whole group of Hubble-Sandage variables to which the two stars belong. Thus the results obtained from the study of these variables confirm our suspicion that we should not classify a star as a supernova simply because it satisfies Zwicky's two conditions. Moreover, these results provide a strong inducement to investigate further this new class of variables, which almost certainly includes Eta Carinae and other well-known objects.

The *Second Supplement to the Third Edition of the General Catalogue of Variable Stars,* published in 1974, did in fact introduce a new class of variables which included the objects we have discussed. S Doradus was chosen as the prototype of this class; but Eta Carinae was *not* included. After all we have said, this might appear a bit paradoxical, but one should add that most of the findings linking Eta Carinae to the Hubble-Sandage variables came to light after the *Supplement* was published. It is also pos-

sible that these high-luminosity blue variables do not constitute a homogeneous group but will have to be redistributed into separate classes. Be
that as it may, for the time being it would seem advisable not to include
in well-understood classes of variables, like novae and supernovae, objects that might be of an entirely different type. For instance, if a variable
does not altogether fit in the new class of S Doradus stars, it may be better to set it aside with the harmless label of "peculiar." On the whole,
perhaps, it might also have been better to wait a while longer before introducing the whole new class of S Doradus stars.

CONCLUSION

We began this discussion in an effort to solve the mystery of a star that
appeared unique in the whole universe. After gathering and examining
the contributions made by the many astronomers who have studied it for
three centuries, we have been able, not without some effort, to find a
convincing interpretation. Future research will tell if it is the right one.

On the whole, the study of Eta Carinae has had very far-reaching results. Along this journey of discovery we have found at least one new
class of celestial objects, possibly more. The task ahead of us is to extend
and refine the research in this field which promises to give us new insight
into particular phases of stellar evolution. Unfortunately, the study of
the variables we have discussed in this chapter, including Eta Carinae, is
hampered by the fact that they evolve very slowly, and therefore observations spanning several generations are required in order to understand them. Furthermore, they are so rare that very few objects of this
type have been found, and none of them is close enough to permit detailed studies.

To compensate for these difficulties, there is the fact that they are very
luminous. Thus we can predict that great progress will be made in the
years to come, and that the world's largest telescopes will be able to probe
some of their secrets, at least in the nearest galaxies.

Even so, this research will require a very long time. Generations of
astronomers will have to work solely to help the next one. Their only
reward will be the modest satisfaction of having made a partial contribu-

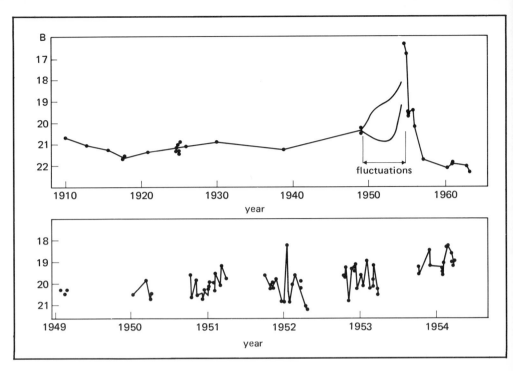

Figure 4.17 Light curve of the variable V 12 in the galaxy NGC 2403. The magnitudes are given in the B system. The fluctuations exhibited by the star during its rise to maximum are reproduced on an expanded scale in the lower portion of the picture. (From *The Astrophysical Journal.*)

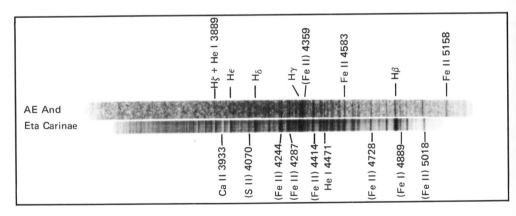

Figure 4.18 A comparison of the spectra of AE And and Eta Carinae. Note the large number of emission lines (which appear dark because the spectra are reproduced in the negative) common to the spectra of both stars. (Courtesy of R. Humphreys.)

tion to the whole. Sometimes, they will not feel even this much satisfaction, since the stars they spend their lives observing might not yield their secrets until a much later time. Nevertheless, they will be sustained throughout by the knowledge that all their work will become meaningful when the sum total of many individual efforts finally yields the solution to the problem—even if that is a hundred years away.

5   BLACK HOLES

STRUCTURE AND CHARACTERISTICS

END OF A STAR

As the sun shines in the sky in all its splendor, it is hard to believe that one day, having exhausted its nuclear fuel, it will be forced to shut off. Yet, just like us and all that surrounds us, the sun will also die. The fact that its lifetime is measured in billions of years, while ours spans at most a century, does not change the harsh law of nature that sooner or later everything must come to an end.

Thinking of that distant time when not only mankind but most of what we see in the universe will either no longer exist or be altered beyond recognition, we may wonder how the sun will end, what it will be transformed into. On the basis of our current knowledge, we can make a fairly accurate prediction: after producing energy for about 5 billion years by "burning" hydrogen into helium in its interior, the sun's supply of hydrogen will be exhausted. The conversion of helium into carbon and oxygen will follow, and these two, in turn, will be burned into yet heavier elements. The formation of these elements generates ever smaller amounts of energy.[1] Eventually, when all the lighter, energy-producing elements have been depleted, energy will no longer be generated in the interior of the sun. In the absence of the internal pressure that supported them, the outer shells will rapidly fall toward the center due to gravitational attraction. In the course of this very rapid collapse, the atoms will be squeezed together ever more tightly, and the electrons will be disassociated from the nuclei. This process will continue until the pressure of the electrons becomes sufficiently high to halt the contraction and to maintain the entire stellar mass in equilibrium. At this point the sun will be reduced to the size of the earth; its matter will be so tightly compressed that a tablespoon of it will weigh 1,000 tons. Under these conditions, the sun will emit a whitish light, and will be 10,000 times fainter than it is now. In other words, it will have become a "white dwarf."[2]

1. See *Beyond the Moon*, p. 24.
2. See *Beyond the Moon*, pp. 108–110.

One might think that all stars end in this manner, by becoming white dwarfs, but it is not so. As early as 1931 S. Chandrasekhar had shown that no white dwarfs can exist with masses in excess of a specific value. According to the most recent computations, this value is 1.25 solar masses. For greater values of the mass, the collapse cannot be stopped until most of the free electrons have been squeezed into the positively charged nuclei, thereby transforming most of the star's matter into neutrons. Only at this point do nuclear forces between particles become strong enough to halt the collapse. The resulting object will be much smaller and much more dense than a white dwarf. At most, it will have a diameter of 30 km and such a high density that one tablespoon of matter sampled from its interior will weigh 10 billion tons. Such a body has been named "a neutron star."[3] The existence of these stars was predicted theoretically by R. J. Oppenheimer in 1939, and it appears confirmed by the discovery of pulsars and by the studies of the Crab nebula. However, in the late 1960s it was demonstrated that neutron stars cannot form from stars with arbitrarily large masses. Assuming that general relativity holds, and that the velocity of sound in the material of which neutron stars are made is less than the speed of light, it was shown that a neutron star cannot have a mass larger than about 3 $M_\odot$.

This conclusion is of great significance. In fact we know of many stars with masses as large as 20 to 30 times that of the sun. It is natural therefore to ask how such stars end their lives, since they are so massive that they can neither end as white dwarfs, as we believe will happen with the sun, nor as neutron stars. One could envisage two possibilities: either the massive star, before exhausting its nuclear fuel and undergoing collapse, sheds so much matter that its mass is reduced to below 3 $M_\odot$ or the collapse is not stopped. The phenomenon of mass loss is certainly observed in many stars, but no specific process is known through which a very massive star can reduce its mass below a critical limit prior to exhausting its nuclear fuel. Theoretical astrophysicists therefore believe that stellar collapses are more likely to occur in objects with initial masses greater

3. See *Beyond the Moon*, pp. 157–163.

than 3 $\mathcal{M}_\odot$. They maintain that the gravitational collapse cannot be stopped under these conditions; so the star will continue to contract until it is reduced to a point.

THE BLACK HOLE

The process of infall is not only hard to imagine but has extraordinary consequences. The most important of them is the ever-increasing curvature of the space that surrounds the collapsing star. To understand this effect better, we need to recall briefly how the concept of mass is introduced by A. Einstein in the *Theory of General Relativity*.[4] According to Einstein, the gravitational mass of a body is manifested as a geometric property of the surrounding space. To illustrate this concept, let us imagine that we can represent 3-dimensional space with a rubber sheet. In the absence of masses, the plane will be perfectly flat and smooth, and the lines we can trace on it with a straight edge will be straight lines (figure 5.1). Let us assume now that we place on this plane a large ball weighing 10 kg. The rubber sheet will be slightly bent under its weight, and the lines we have traced will appear to follow a path whose curvature increases the closer they are to the center of mass of the ball. At a distance far enough away from the ball, the lines will remain straight, and the plane will appear as flat as it was everywhere before we placed the ball on it. Suppose now that we compress the 10-kg sphere into a very small ball still weighing 10 kg. If we place it on the rubber sheet, the sheet will bend within a much smaller region but yield a much greater curvature; so the plane in that region will assume the shape of a drop. This analogy illustrates the curvature of space produced by the presence of a mass according to general relativity.

One of the main observational proofs of general relativity is the detection of the curvature of space in the proximity of massive bodies. The technique used is conceptually very simple. Suppose that we observe the position of a star on a given night and measure it again when the sun is

---

4. The mass of a body can be (1) the mass of a moving body in a gravitational field, (2) the mass that produces a gravitational field. In this chapter we refer to the latter, called "the active gravitational mass."

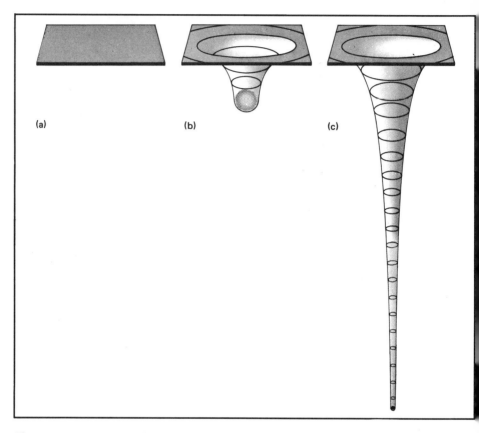

Figure 5.1 (a) A two-dimensional flat space (represented by a rubber sheet) in the absence of a mass. (b) The same space curved by the presence of a slightly dense mass. (c) The same space warped into an elongated funnel by a concentrated mass.

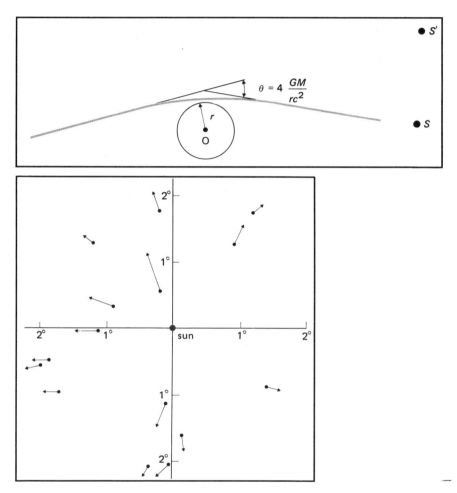

Figure 5.2 Deflection of light in the gravitational field of the sun. *Above*, a schematic illustration of the phenomenon; the star which is located in *S* appears instead in *S'*. *Below*, displacements of the stars in the vicinity of the sun as observed by Campbell and Trumpler during the total eclipse of September 21, 1922.

near our line of sight to the star (figure 5.2). Adopting Einstein's premise that the mass of the sun curves space, the light coming to us from a star should be bent with the same curvature. The star would then appear in S' rather than in S. Measurements of such deviations are possible when the solar disk is occulted by the moon during total eclipses of the sun. By comparing photographs of a stellar field obtained during an eclipse with others taken when the sun is no longer in the region, we should observe that in the first case the stars appear slightly farther away from the location in the field corresponding to the center of the sun. The closer the stars are to this point, the larger this displacement should be. This observation has been attempted repeatedly for over half a century, but only recently could definite conclusions be reached by using radio sources rather than optical ones such as stars.

Returning now to the discussion of the large mass during collapse, it follows that as long as the mass is distributed in a large volume, space will be curved very little; but as the infall progresses, the ever-increasing concentration of mass into a point will curve space more and more. As the star keeps contracting ever more rapidly in the interior of this well, the pressure, gravitational field, and density tend to become infinite, and the collapsing matter approaches a singularity—a point outside our space-time.

Again representing 3-dimensional space with a 2-dimensional plane (as in figure 5.1), we can reproduce the collapse of a star by imagining that we concentrate the 10-kg ball into an ever-decreasing sphere. This will stretch the "drop shape" (figure 5.1) until it is transformed into a funnel (figure 5.3). The transformation of the drop into a funnel represents the deepening of the potential well in the space surrounding the star until the walls of the well become so steep that they are almost parallel, and thus form an opening through which we can break out of the universe.

For now we will not worry about the meaning of the phrase "breaking out of the universe." We will return to this subject later. But at this point we want to note a more dramatic aspect of the phenomenon: the transformation of the drop into a funnel means that in effect the well no longer has a bottom from which an object can work its way back up. In

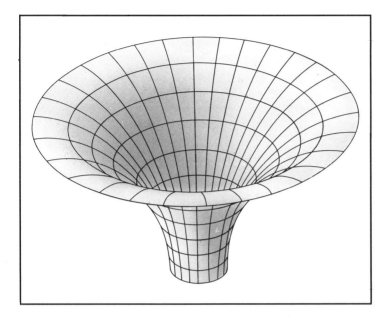

Figure 5.3   The curvature of space in the presence of a strong gravitational field. The two-dimensional space shown in figure 5.1 is warped into a funnel at the bottom of which is the collapsed object that generates the field. (From K. S. Thorne, *Scientific American*, December 1974.)

other words, the star which is indefinitely collapsing warps the surrounding space in such a way that an object that may chance to fall into the funnel can no longer climb up to the plane, or to the region of space where the attraction of the collapsing mass is practically negligible.

To discover additional characteristics of the strange structure that forms during the collapse of the star, let us follow it, imagining we find ourselves on the surface of the star at the onset. Let us suppose that we have available a powerful searchlight with which we can send luminous signals. Before the collapse starts, we can send signals in any direction, but from a certain moment on during free fall toward the center of the star things will no longer be that simple. If we point the searchlight vertically, the light will leave the star in a straight line; but if we place the searchlight at an angle, the rays of light will be bent because of gravity. When the angle between the searchlight and the surface of the star is small enough, they will fall back onto the star (figure 5.4). The limiting value of the angle—below which we cannot aim the searchlight if we wish its light to leave the star—defines a cone called the "escape cone." As the star continues to collapse, space-time in its vicinity will bend more and more, and the exit cone will become ever smaller, until the searchlight can send luminous signals only vertically. All other signals, in whatever direction they are sent, will fall back onto the star.

We have noted that the collapse cannot be halted. As it proceeds, the gravitational field becomes so intense that eventually even the beam sent vertically will be captured, because the escape velocity (that is, the value of the velocity the beam should have to escape from the star) will have become greater than the speed of light.

From this point on, light can no longer escape the star, and we can say that we have reached the "horizon of events" or simply the "horizon." We can no longer communicate with the external universe, and we, in turn, will have disappeared from it. In place of the star there will be left in space a region some kilometers in diameter which, being no longer able to send light to the outside, will appear black. The star has now vanished from our universe and has been transformed into a "black hole."

What we have described, by imagining ourselves on the surface of the collapsing star, has occurred in a time not only finite but extremely short,

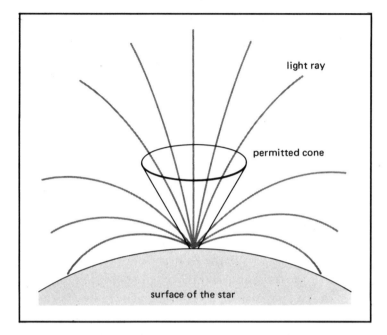

Figure 5.4   Effects of a gravitational field on the rays of light emitted by a searchlight on the surface of a collapsing star. Only those emitted within the cone can escape; all the others fall back onto the star because of the very strong gravitational field.

as we ourselves could have observed by means of a watch. Things would have appeared very different, however, to somebody observing the same phenomenon from the outside. Given that according to general relativity time slows down in a gravitational field, the external observer would have noted that our clock was becoming ever slower as the intensity of the field increased and then stopped altogether upon reaching the event horizon. Actually, the external observer would have never seen us reaching the horizon. While he was watching our clock slow down, he was also observing us traveling ever more slowly, and he would have had to wait an infinite time to see us reach the event horizon. One should remember at this point that the atoms from which light originates also appear to the external observer as tiny clocks that move slower and slower, so that the frequency of their emitted radiation becomes ever lower. Thus, observing the gravitational collapse from the outside, we will see that the light emitted by the collapsing star is becoming ever fainter due to the narrowing of the escape cone and ever redder due to the slowing down of time. Both phenomena are due to the fact that the gravitational field is becoming stronger near the star's surface, as the latter becomes increasingly smaller.

The diameter of a black hole, that region of space cut out from our universe and enclosed by the event horizon, depends on the infallen mass and is generally quite small. Should we be able to compress the entire earth into a black hole (something we could only achieve artificially by squeezing it in a giant vise and furnishing a lot of energy), we would obtain an object with a diameter of only 18 mm. A black hole formed with the entire mass of the sun would be no larger than 6 km. Very massive stars, those in fact that should collapse into black holes, would form holes with diameters between 10 and 150 km.

THE SUICIDAL ASTRONAUT

A little while back we imagined finding ourselves on the surface of a collapsing star. Let us suppose now that a daredevil astronaut really wishes to live this adventure, by placing himself on the surface of the collapsing body. Although he is on a very massive body, he will not be crushed by his own weight as would happen if he found himself on the surface of a neutron star or a white dwarf. On the contrary, since he is in free fall, he will feel no weight, as would happen inside an elevator whose cables had snapped. Unfortunately, however, he will experience the effects of gravitation as tidal forces which will rapidly increase across his body until, in a limited and relatively short period of his subjective time, they reach an infinite value.

To better understand this effect, let us recall the tides produced by the moon on our oceans. For simplicity, let us assume that earth is completely covered by the sea. In the absence of external bodies, the water would be distributed on the surface in a layer of uniform thickness (figure 5.5). The moon, however, exerts its gravitational attraction on our entire planet which will tend to move toward it. Newton's gravitational law tells us that the attracting force is not only directly proportional to the mass but also inversely proportional to the square of the distance. This means that, for equal masses, a body will be less attracted to another the more distant it is. If we consider separately the points M_1, T, and M_2, we can see that M_1 will be subject to a greater force and will move toward the moon more than T. T, in turn, will tend to approach the moon more than M_2. In other words, the water on the side facing the moon will be lifted toward it, drawing some from the sides; the land under it will also move toward the moon but a little less; and the water on the opposite side will lag behind. Therefore in M_1 and M_2 there will be high tides, while on the sides there will be low tides. A similar phenomenon is experienced by our bodies as well: when we stand on the surface of the earth, our feet are attracted toward the center more than our heads. The difference between the two forces is, however, extremely small because the distance from head to foot (roughly 2 meters for the height of a man) is very small compared to our distance from the center of the earth (6 million meters).

The case of the astronaut falling into a black hole is altogether different, since in this case a mass at least a million times greater than that of earth is concentrated in an extremely small region. In these circumstances, when the star has become sufficiently small, the distance between his feet and head will be very large as compared to that from his feet to the center of the collapsing star. Head and feet will therefore be subjected to greatly differing forces, and the body of the astronaut will be stretched more and more until it is horribly torn apart, even before falling into the singularity where finally he will be annihilated.

This disastrous end would occur rapidly only while exploring a black hole of a few solar masses. The tidal effects that occur at the event horizon are inversely proportional to the square of the mass of the hole. Should the black hole have a mass of 100 million $M_\odot$, these effects would become even smaller than the already negligible ones produced by gravity on the surface of the earth. In this case the astronaut could cross the event horizon unharmed and survive for a few minutes within a black hole. If the mass of the black hole were as large as 10 billion solar masses, he could survive for a whole day. But eventually he will end up in the same manner—torn apart by tidal forces and then crushed and swallowed by the singularity.

TYPES OF BLACK HOLES

The black hole we have been discussing up to now is that of the ideal case predicted by K. Schwarzschild in 1916 and further studied by J. R. Oppenheimer and H. Snyder in 1939. The last authors, more precisely, considered the gravitational collapse of a nonrotating, spherically symmetrical star. A real star is certainly not as simple. Rotation appears to be a general phenomenon in the whole universe; there is no body, or system of bodies that, after being carefully investigated from this viewpoint, has not shown rotation. In particular, we do not know of any nonrotating star. Therefore the progenitors of the black holes, stars more massive than 3 $M_\odot$, must also be rotating. In addition, like most known celestial bodies, they should exhibit irregularities, differ from spherical symmetry, and almost certainly be endowed with magnetic fields.

Thus there should exist black holes originating from the collapse of stars with these characteristics. Theoretical astrophysicists maintain that they will be different from the type of black hole examined up to now and different among themselves depending on whether they possess one or another or all of the properties mentioned above. The theoretical study of this subject started in 1963, when R. P. Kerr first considered rotation, and was further developed after 1965, the year in which E. T.

Newmann found a new solution to Einstein's equations corresponding to the so-called Kerr-Newmann black hole.

At present at least four different types of black holes are believed possible. They differ from each other depending on the parameters used to characterize them. They are:

1. Schwarzschild black holes, the type discussed so far, which are non-rotating and characterized only by their mass;
2. Reissner-Nordstrom black holes, nonrotating but endowed with mass and electrical charge;
3. Kerr black holes, endowed with mass and rotation but no electrical charge;
4. Kerr-Newman black holes, characterized by mass, rotation, and electrical charge.

The black holes of the last three types are extremely interesting, not only because they are presumably closer to reality but also because they possess a further very important property. They are endowed with an ergosphere,[5] a zone outside the event horizon in which one can conceive of processes for extracting energy from the black hole. One such process, first proposed by R. Penrose in 1969, is conceptually quite simple. If, as shown in figure 5.6, a particle, P_0, falls into the ergosphere from infinity, it might split in two in such a way that one component, P_2, falls into the black hole with negative mass-energy, while the other, P_1, escapes to the outside with a mass-energy greater than that of the initial particle, P_0. Such energy is obtained at the expense of the rotational or electromagnetic energy of the black hole. In this manner, as a very large number of particles traverses the ergosphere, an enormous amount of energy could be extracted from the immediate surroundings of the black hole gradually slowing its rotation, and, finally, the black hole would stop altogether. Similarly, particles crossing the ergosphere could extract electromagnetic energy.

Almost certainly, at the moment of its formation from the gravitational collapse of a massive rotating star, the black hole is of the Kerr-

5. This term was coined in 1970 by R. Ruffini and J. A. Wheeler.

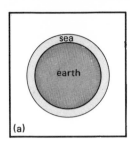 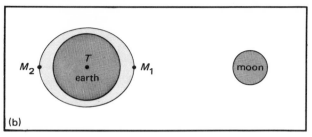

Figure 5.5 Tidal effects: (a) uniform distribution of the waters over the surface of the earth in the absence of external forces; (b) displacement of the liquid masses (tides) in the presence of the moon.

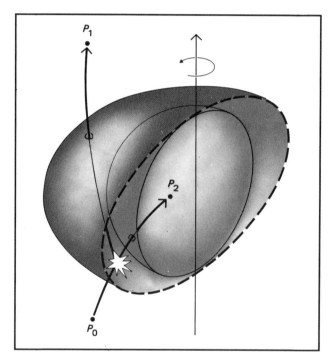

Figure 5.6 Representation of a section of space around a black hole. The trajectory of a particle which extracts energy from the ergosphere is shown. The ergosphere is the region between the dashed boundary line and the event horizon (solid line). If a particle breaks apart in this region, and one of the fragments falls into the black hole, the other fragment can escape to infinity with a kinetic energy greater than that of the original particle. The difference is energy subtracted from the black hole. (From *Physics Today*.)

Newmann type. Through the processes just mentioned, the black hole
can lose its angular momentum, that is, its rotation, and decay into
a Reissner-Nordstrom black hole. Alternatively, its charge could be neu-
tralized, and it could be transformed into a Kerr black hole endowed
only with mass and rotation. In both cases there is still an ergosphere
from which energy can be extracted. Following additional loss of charge
or angular momentum, these black holes can be reduced to one of the
Schwarzschild type, which lacks an ergosphere and from which energy
can no longer be extracted.

The mechanisms by which a charged or uncharged particle can extract
energy from the ergosphere are still not clear. Even harder has been the
theoretical or observational search for physical processes that could actually be operating in nature. On the other hand, a formula has been
found to measure the amount of energy that can be extracted from
the ergosphere of a black hole. Depending on the type of black hole, this
amount differs but is always enormous. In the conversion from a
Reissner-Nordstrom black hole to a Schwarzschild black hole, the extracted energy can be as high as 50 percent of the total mass-energy.
Considering that in order to produce the enormous amount of energy
radiated by a star into space only a minute percentage of its rest mass
is used, one can understand the important role that black holes could
play in the general economy of the universe.

These theoretical results, obtained since 1971, have fundamentally
changed our viewpoint on the subject of black holes. They appear to us
today not only as bottomless pits from which infallen matter and energy
can never return but also as the greatest reservoirs of energy in the
universe.

This possibility has so stimulated the imagination of physicists as to
inspire speculations about practical utilization of the energy extractable
from a black hole. C. W. Misner, K. S. Thorne, and J. A. Wheeler light-
ened the tone in their massive book, *Gravitation*, with a description of a
vast city built by an advanced civilization on a rigid framework surrounding a black hole (figure 5.7). Every day trucks carry tons of garbage out of the city to a dumping ground. The garbage is loaded into
shuttle vehicles which are then sent, one after the other, toward the black

hole. Each vehicle enters the ergosphere, passes close to the event horizon and, upon reaching a certain "ejection point," dumps its cargo into the black hole. Theory shows that by properly selecting the orbit of the vehicle, the load of garbage can be ejected with negative energy-at-infinity. As the garbage falls down the hole, changing the hole's total mass-energy by

$$\Delta M = E_{\text{garbage ejected}} < 0,$$

the shuttle vehicle recoils from the ejection point with more energy-at-infinity than it previously had.

The vehicle deposits its huge kinetic energy in a giant flywheel adjacent to the garbage dump. The flywheel rotates, depositing the vehicle where it can be reutilized for another load, while at the same time powering a generator (for example, a dynamo) which produces electricity for the city. Thus for each trip of the vehicle a certain amount of electrical energy is generated, which can be computed as follows:

Energy per trip
$$= E_{\text{vehicle up}} - (\text{rest mass of vehicle})$$
$$= (E_{\text{vehicle + garbage down}} - E_{\text{garbage ejected}}) - (\text{rest mass of vehicle})$$
$$= (\text{rest mass of vehicle} + \text{rest mass of garbage} - \Delta M) - (\text{rest mass of vehicle})$$
$$= (\text{rest mass of garbage}) + |\Delta M|.$$

Recalling that ΔM represents the amount by which the mass-energy of the black hole decreases, it follows that the inhabitants of the city can use the black hole for many purposes. Not only can they eliminate their garbage and convert its entire rest mass into kinetic energy of the vehicle, and then into electrical power, but they can also convert into electrical power some of the mass of the black hole itself with a conversion efficiency close to the 50 percent limit.

It is obvious that the practical realization of this undertaking would be fraught with many and severe difficulties, such as the problem of transforming the kinetic energy of the returning vehicles into electrical power without destroying the flywheel, or the problem, which at first sight seems trivial, of finding enough garbage to discharge into the black hole

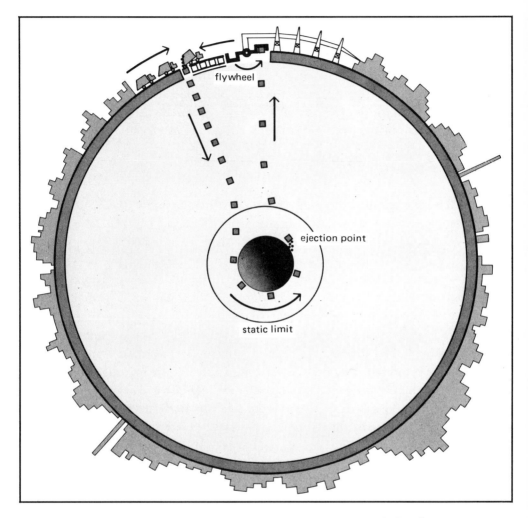

Figure 5.7 A space city built on a rigid framework around a black hole. Electrical energy is derived from the dumping of garbage into it, according to the concept of ergosphere. (From C. W. Misner, K. S. Thorne, and J. A. Wheeler, *Gravitation*. Freeman, © 1973.)

without destroying the city. The latter effort might, in fact, become quite serious since all the materials brought from the planet of origin should be recyclable. According to D. L. Block, who discussed this subject in the magazine *Sky and Telescope*, the greatest difficulty would reside in building the giant sphere around the black hole. It would be a titanic enterprise even for our hypothetical "super civilization." But at the moment, perhaps, an even greater difficulty is to *find a black hole!*

SEARCH FOR BLACK HOLES

The preceding paragraph concluded with a statement that may appear too pessimistic and disappointing. After all, we have just become acquainted with extraordinary objects. They are holes in the fabric of our space-time. They are invisible, but they are endowed with mass which continues to exert an influence on nearby bodies. They are capable of swallowing and capturing from our universe things or living beings unfortunate enough to approach it too close, but at the same time they are charged with so much energy as to make the huge amounts radiated into space by stars appear negligible. These objects, these black holes, have been proposed and described by theorists who have investigated their birth and evolution. Their research has given us an overall view so new and unusual as to convince us that our entire body of knowledge on the universe could be revolutionized in the very near future. However, taken by this exciting adventure of the mind, we have not yet asked the question, Do black holes really exist?

The uncertainty we present is not intended to belittle the results obtained by great scientists but to make us consider the possibility that all the marvelous discoveries and speculations on this subject could be the result of a purely theoretical construction which may not exist in the real world. It is now time to explore the sky and attempt the actual discovery of black holes.

At first sight, this quest appears hopeless. How can we hope to find a body which cannot be seen in its own light and could not be seen even if illuminated by other sources whose light it would swallow? The only possibility would seem to be to observe it projected against a luminous source. But even this method seems impractical if we consider how seldom such an alignment may take place, especially in view of the small size of black holes and the very great distance at which even the nearest one could be found. We can see Venus or Mercury very well when they transit the sun, but, if we reduced them to black holes, their diameters would shrink to a few millimeters, and we could no longer see them. On the other hand, for black holes to have diameters of just a few tens of kilometers, they should have masses several times greater than that of the sun, which would place them outside the solar system. But even if these black holes were as close as Venus or Mercury, they would still be too small to be detected when transiting the sun.

Despite these meagre prospects, theorists have felt from the very beginning of their work the need to confirm observationally their predictions and have searched for some way to discover black holes, at least indirectly. The first method that comes to mind is to reveal the existence of a black hole through the only effect still present outside the event horizon, namely, its gravitational attraction. If it were possible to squeeze the sun into a black hole a few kilometers in diameter, darkness would descend on the entire solar system. The earth and the other planets would be illuminated only by starlight, and we would no longer see the sun, not even as a black disk projected against the starry sky, which would be the case if it ceased to be luminous but maintained its current size. The planets would continue to circle it as they do now, since its mass, even though invisible, would remain the same, and the gravitational attraction exerted on the earth and the other bodies would not change in the least. An astronaut coming from another planetary system and discovering the planets in our system could deduce the presence of the sun from their motion and compute its mass, even though the sun was now a black hole.

In a similar manner, we could use the stars to search for possible black hole members of binary systems. It is well known that there exist many binary systems, in which two stars rotate about their common center of mass. Even when they are so close that they appear to us as a single star, the binary nature of the system can still be revealed by the reciprocal periodic eclipses or by the oscillations toward higher and lower frequencies of the two groups of spectral lines emitted by the two stars. From these oscillations both the mass of the stars and certain orbital parameters can be computed.[6] If one of the two components is much fainter than the other, we cannot observe its spectrum, rather we will observe oscillations to higher and lower frequencies of all the lines of a single spectrum. Even in this case, we can still deduce the mass of the system. Then, if we can estimate the mass of the visible star from its spectral type, we can obtain the mass of the invisible companion by subtracting the former from the total. Should the invisible body in the binary system turn out to have a mass greater than 3 $\mathcal{M}_\odot$, we would have good reason to believe that we have discovered a black hole.

Following this approach, in 1964 Soviet astronomers Y. B. Zel'dovich and O. K. Guseynov began the search for black holes by selecting from catalogues of spectroscopic binaries those in which only a single spectrum was observed, and among these the few more massive than the sun. The investigation resulted in five possible candidates. Four years later, V. Trimble and K. S. Thorne revised and expanded the list of the Russians, finding a total of eight potential candidates. However, Trimble herself played the devil's advocate, reviewing critically each of those cases and finding a reasonable explanation for all of them without invoking black holes.

In fact, there might be several reasons why the second object could be invisible without being a black hole. For instance, the dark companion could appear to have a large mass only because it is in turn a multiple system of stars, all rather faint; or it might not really have a large mass because the mass of the luminous component had been underestimated. After these disheartening conclusions, enthusiasm for the black hole search seemed to be on the wane. Interest was rekindled, however, only two years later by a new method which was almost the opposite of the first.

As early as 1964, several theorists had suggested that a black hole in a binary system can capture gas from the luminous companion. This gas, rushing into the hole, can be heated to temperatures high enough to result in X-ray emission (figure 5.8). Therefore, if one could find a binary

6. See *Beyond the Moon*, pp. 130–134.

system in which the dark companion was very massive and emitted X rays, one had reason to consider the dark object a good candidate for a black hole. Following this chain of inference, it was easier to conduct the search in a manner contrary to the former method, namely, to discover X-ray sources, search for an optical counterpart for each of them, and determine, in case it turned out to be a star, if it had the spectroscopic characteristics of a component of a binary system.

In order to observe possible X-ray emission, it was necessary to send instrumentation above the earth's atmosphere, which, luckily for us, prevents this dangerous radiation from reaching the ground. The search for X-ray sources was first systematically pursued by R. Giacconi and his colleagues in 1971 by means of the Uhuru satellite, launched December 12, 1970, from the San Marco platform. This satellite, through an all-sky scan, discovered over one hundred new X-ray sources, and most important, during its long life in orbit was capable of monitoring for long periods some of the more interesting ones.

First of all, it was found that none of the eight candidate objects previously mentioned was an X-ray source. However, it was also discovered that six intense X-ray sources were certainly associated with binary systems. Five of them showed a periodic disappearance of the X-ray emission, which could only be interpreted as the periodic occultation of the X-ray source by another object (later identified with a visible companion in all five cases). The sixth source did not show X-ray eclipses, although it later became apparent that it was perhaps the most interesting of all. The six sources were Cen X-3, Her X-1, 3U 1700-37, Vela X-1, SMC X-1, and Cyg X-1. The first two exhibited X-ray pulsations with respective periods of 4.84239 and 1.23782 seconds and therefore correspond to neutron stars. In the other four the optical counterparts were found to be supergiant stars, which, as was learned from spectral measurements, exhibited gas loss. The spectrum of each of these supergiants exhibited lines whose frequency oscillated periodically. The measurement of the amplitude of the oscillations allowed determination of the mass of each system. X-ray emission was presumably due to the heating of gas lost by the luminous star while falling onto the dark companion through its exceedingly strong gravitational field. This field could be produced by a white dwarf, by a neutron star, or by a black hole. To distinguish between these three cases, it was necessary to obtain the mass of the dark companion. In the case of 3U 1700-37, Vela X-1, and SMC X-1, the masses were found to be 2.5, 3, and 2 $\mathcal{M}_\odot$, respectively. In the case of Cyg X-1 the mass turned out to be about 8 $\mathcal{M}_\odot$. Taking into account the limit of 3 $\mathcal{M}_\odot$, above which a collapsing star cannot reach a white dwarf or a neutron star configuration, one could easily conclude that at least Cyg X-1 (figure 5.9) was a black hole.

Let us now consider in detail this very interesting object. The source Cyg X-1 had been discovered in 1964, that is, much before the launch of the Uhuru satellite. The instruments onboard this satellite, however, recorded variations in the emitted intensity in time scales of the order of one-tenth of a second. The rapidity of these variations already suggested that the emitting object was very small. It would have been interesting to discover if the X-ray source coincided with an optically visible object. Unfortunately, at that time, the position of X-ray sources could not be determined with the same precision as that of optical or radio objects. The region defining the position of the X-ray source was always large enough to contain several stars. Fortunately, in early 1971 the Westerbork radio telescope discovered a weak radio source in the region of Cyg X-1. The source had to be of variable intensity, since previous scanning of the region with radio telescopes had revealed no radio source. Once its position had been exactly determined, it was found to coincide with that of a catalogued blue star, HDE 226 868. Sometime later, checking the observations made by Uhuru over a 1-year period and comparing them with radio measurements, H. D. Tananbaum found that in March 1971 the X-ray emission had undergone a sudden transition concurrent with the onset of radio emission. This phenomenon has never been fully explained, but it gave strong indication that the X-ray source, the radio source, and the visible star were all the same object. This object could only be a binary system in which the invisible component was the X-ray emitter and the other component was the visible star. Confirmation came immediately afterward with the discovery that the latter was one of the

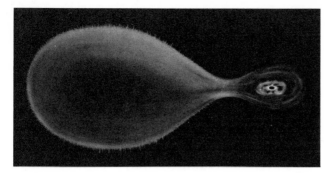

Figure 5.8 An artist's conception of a binary system consisting of a giant star and a black hole accreting matter lost by the companion. The infalling material forms a ring around the black hole before being sucked into it. During infall the material is heated by compression, emitting X rays. (From *Sky and Telescope*.)

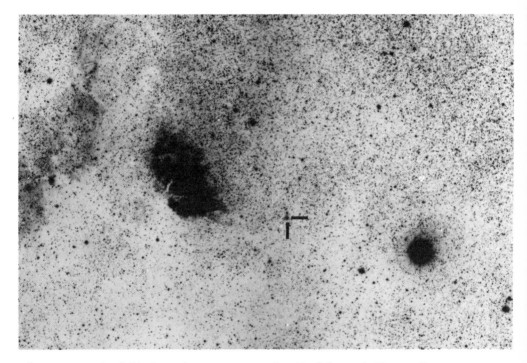

Figure 5.9 The field where the X-ray source Cyg X-1 is located. The two lines indicate its optical counterpart. The star just above it has nothing to do with the system and appears close for reasons of perspective. The region of Cyg X-1 can also be seen at the lower right of the wider-field photograph shown in figure 3.6. (Palomar Sky Survey.)

two components of a spectroscopic binary system, obviously the visible one, revolving around the common center of mass with a period of 5.6 days.

Subsequently the picture became increasingly clearer. Scientists found the star to be a B0 Iab supergiant; discovered an emission line of helium indicative of a gas stream (already predicted, as we have seen, for black holes); and were able to determine that the binary system, consisting of the X-ray source and the blue star, revolved in an orbit with a 27° inclination with respect to our line of sight. Now only the most important parameter remained to be determined: the mass of the two bodies. To accomplish this, it was necessary to make sure that the star was indeed a supergiant.

By late summer 1973, on the basis of the orbital data, one could envisage two equally possible solutions:

Characteristics	1st Solution	2nd Solution
Mass of the B star	~30 $\mathcal{M}_\odot$	~1 $\mathcal{M}_\odot$
Luminosity of the B star	~10^5 $\mathscr{L}_\odot$	~10^3–10^4 $\mathscr{L}_\odot$
Mass of the X-ray object	~6 $\mathcal{M}_\odot$	~0.9 $\mathcal{M}_\odot$

Evidently, if the first solution could be proved right, the X-ray object would be a strong candidate for the black hole.

This question could be settled by determining the distance as precisely as possible. The second solution implies a brightness of the B star from 1,000 to 10,000 times that of the sun, while the first one requires a luminosity 100,000 times that of the sun. Since the apparent magnitude of the star is the same in either case, in the first solution the system would have to be farther away from us than in the second. Observational efforts were initiated to solve this problem. In late 1973 two different scientific groups, working independently and using different methods, reached the conclusion that Cyg X-1 was at a distance of about 8,000 light-years. This result corresponded to the greater distance, that is, to solution 1; consequently, the X-ray source, with a mass of 6 $\mathcal{M}_\odot$ (considerably higher than the critical mass of 3 $\mathcal{M}_\odot$) could very well be a black hole.

On second thought, however, one could point out that this result was only a negative proof. It showed that in Cyg X-1 there was a large mass that did not correspond to any known star, which did not necessarily mean that the invisible object was a black hole. It was a very likely possibility but not a certainty.

Additional proofs would have to be found. Again, since it was not possible to look for them in the black hole itself, one would have to search its environs. Theory predicted that in the most likely hypothesis of a rotating black hole accreting material from the visible star the infalling gas would be distributed in a disk, or actually in a vortex in very rapid rotation around the hole. This vortex, located outside the event horizon, would certainly be visible to an external observer. Thus it was still possible to deepen our understanding of black holes by building a plausible theoretical model that would predict its observable properties and then attempting to discover them.

Since 1971 several scientists have studied the structure of a binary system which emits X rays because of the infall of matter from a normal star into a black hole. Almost all of them have been led to the same basic model shown in figure 5.10.

Let us suppose that a black hole is placed at a certain distance from the blue star. Gas lost by the star falls into the hole under its gravitational attraction. As the gas approaches the hole, it winds itself around it along orbits that at first are almost circular; then, because of attrition between adjacent gas filaments, these orbits become spiral, sucking the gas slowly inward. Thus an accretion disk is formed, as shown enlarged in figure 5.10b. The disk has a thickness of about 100,000 km; that is, it is relatively thin because of compression due to the gravitational pull of the black hole located at its center. At the same time, the thermal pressure of the gas tends to counterbalance the gravitational compression and thicken the disk. This occurs mainly in the central regions where the outward pressure of the gas is stronger, owing to the very high temperature produced by the X rays emitted in the immediate vicinity of the black

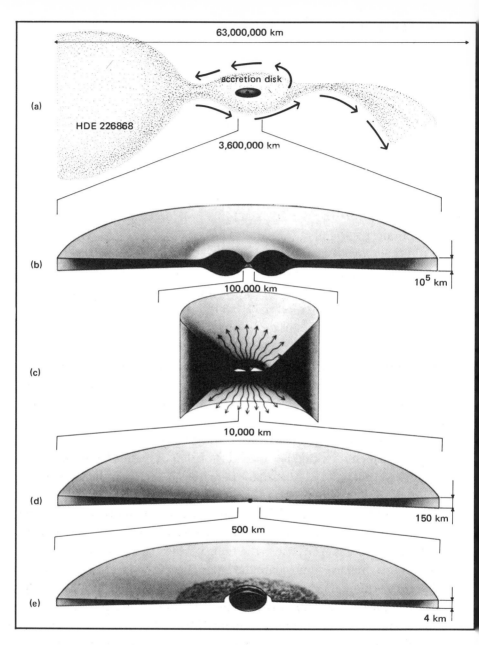

Figure 5.10 A model of the binary system Cyg X-1 shown at increasing magnification (from top to bottom). The monstrous object illustrated in (b) is supposed to be the companion of the supergiant HDE 226868 shown in (a). (From K. S. Thorne, *Scientific American*, December 1974.)

hole. The central region of the accreting disk (figure 5.10c) has a diameter of only 10,000 km, which is smaller than that of earth. Here the pressure of the gas is even higher than in the surrounding bulge; but gravity, having attained a frighteningly large value, keeps the disk very thin, no thicker than 150 km.

After spiraling downward for weeks or months, and getting increasingly closer to the center of the disk, the gas approaches to less than 300 km from it. Here the attraction of the black hole becomes irresistible. In a very short time the gas is sucked away from the disk, crosses the inside horizon, and in a fraction of a second crosses the event horizon and disappears into the black hole. In this inner region of the disk the temperature, which might have been around 25,000° when the gas started to move away from the star, increases to over 10,000,000°, and the gas emits X rays. It is believed that within 200 km of the black hole 80 percent of the emitted energy is in the form of X rays. In the inner 50 to 100 km, as the temperature keeps rising, the disk becomes translucent and very turbulent. At this point the gas is no longer orbiting the hole but is sucked directly into it.

The model we have described, following K. S. Thorne, is consistent with observational evidence, which has disclosed the presence of a gas stream within the binary system. However, the existence of the black hole itself has not yet been proved. The Soviet astrophysicist, R. A. Sunyaev, has suggested a way of making a definite confirmation of it. He believes that in the inner regions of the accretion disk "hot spots" could form, that is, limited and transient patches of much higher temperature than the average which might produce beamed X-ray emission (figure 5.11). In this hypothesis, when the beam strikes the earth (some should come our way at least occasionally), we should observe an X-ray flash at every revolution of the hot spot. The interval between two consecutive flashes should be very brief (a few milliseconds), since the hot spot would be close to the center and circling very rapidly. This phenomenon should last longer than one rotation but no longer than a few minutes.

The X-ray telescope used to detect these regular X-ray pulses must have a large surface area capable of gathering several X-ray photons in

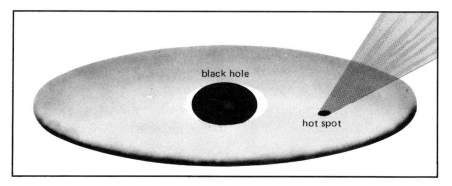

Figure 5.11   A hot spot on the accretion disk of a black hole and its beamed emission. As the hot spot revolves about the hole, its radiation, which is emitted within a cone, sweeps a circle in the sky above the disk. When the beam sweeps by the earth, it can be detected by X-ray counters as an X-ray flare with a pulsed substructure. (From K. S. Thorne, *Scientific American*, December 1974.)

a millisecond. It is estimated that the required surface area should be around 10,000 cm². However, since such large X-ray telescopes have not yet been put into orbit, scientists have used those available to them. As early as October 1973 an X-ray telescope with a surface area of 1,360 cm² and a resolution power of 0.32 ms detected X-ray flashes of about 1/10 of a second, one of which appeared to have a pulsed structure.[7] If phenomena of this type are confirmed, Sunyaev's test will be met, and we will have one more good reason to believe that black holes really exist.

PATHWAYS THROUGH THE UNIVERSE

The possibility that black holes exist lends support to a great number of other theories and hypotheses that have been proposed before, during, and after the experimental results we have discussed. If black holes could be proved to exist, other well-known, but still unexplained, phenomena would appear in a new light, and some theoretical speculations proposed in the last decade could be given serious consideration. Encouraged by the observational results, we will take a decisive step and assume outright that black holes exist. We shall follow the theory to its extreme consequences.

MASSES OF BLACK HOLES
One of the most interesting problems is determining masses. The black holes we have considered have masses of the same order as those of the heaviest stars; that is, from 3 to 20 or 30 $M_\odot$. But it is reasonable to ask if there might be black holes with either smaller or larger masses and, if such were the case, if observational evidence could be sought.

Let us start from very large masses. According to A. M. Wolfe and G. R. Burbidge, many elliptical galaxies harbor at their center huge black holes with masses ranging from 1 billion to 10 billion $M_\odot$. The mass may be concentrated in a single black hole, or distributed among a great number of smaller ones, each of about 1,000 $M_\odot$. This conclusion is

7. R. E. Rothschild, E. A. Boldt, S. Holt, and P. J. Serlemitsos.—Trans.

based on two facts: one is the extraordinarily violent activity observed in the nucleus of certain galaxies, such as the celebrated M 87;[8] the other is the high value of the mass-to-light ratio found in this type of galaxy.

We will stop here a moment to clarify the second point. As we know, there is a relation between the mass of a star and its luminosity. For instance, main sequence stars are heavier and more luminous. Given the distance and the apparent magnitude of a galaxy, we can determine its absolute magnitude, and from this its luminosity (that is, how much brighter than the sun that galaxy is). Furthermore, once we have computed its rotational velocity from spectroscopic measurements, we can determine the value of the total mass, expressed as usual in solar masses.[9] The mass-to-light ratio thus obtained can give us, although indirectly, very useful information. For instance, if all the stars in a galaxy were like the sun, the value of the mass/light ratio would be 1. This value would be the same if the galaxy contained stars both heavier and lighter than the sun, provided their respective luminosities were higher or lower in the proper proportions. The ratio would not be equal to 1, however, if in addition to the luminous stars there should be a number of dark stars that contribute to the total mass but not to the luminosity. Similarly, the value would be different if there were stars as luminous as the sun but with different masses.

For a number of spiral galaxies, such as ours, the determination of the mass/light ratio has resulted in very low values, almost always less than 10. But in elliptical galaxies, this value has turned out to be considerably higher, about 70. According to Wolfe and Burbidge, this anomaly is due at least in part to the presence of one or more black holes in the nucleus of galaxies of this type.

Masses from 1 billion to 10 billion times that of the sun, such as those attributed by these authors to the black holes at the center of giant elliptical galaxies, are of the same order of magnitude as the masses of smaller galaxies. Therefore we should not be surprised to find that besides galaxies space is also dotted with black holes of far smaller size but

8. See *Beyond the Moon*, pp. 229–236.
9. See *Beyond the Moon*, pp. 226–229.

equally enormous mass. Their existence could be deduced from this simple reasoning. Let us assume that there are black holes at the center of certain galaxies. Since a black hole cannot be shut off, each one of them will be the only survivor of its own galaxy, whether it has swallowed all this matter, or just some part of it, while the rest of the galaxy was dispersing into space. However, as S. van den Bergh pointed out, the number of these supermassive intergalactic black holes cannot exceed that of the known galaxies, otherwise the latter would show effects of tidal distortion which in fact have never been observed.

In addition to supermassive black holes, it has been suggested that there might also be mini-holes with masses considerably smaller than that of the sun. At present no mechanism is known that might produce black holes of small mass, but it is conceivable that due to special conditions this could have happened in the earliest phase of the formation of the universe, that is, in the first millisecond of the Big Bang. These mini-holes would have survived until now and would be scattered in enormous numbers throughout the universe.

According to S. Hawking, the smallest black hole has a mass of 10 billion tons, or 1,000 times smaller than that of the minor planets, such as Eros. This mass is concentrated in a sphere a fraction of a millimeter in diameter. In early 1975 G. F. Chapline suggested that the existence of these mini-holes could be proved by background γ-ray emission of about 30 MeV. In this hypothesis, he concludes that in every galaxy there should be 10^{23} mini-holes. Thus they would be 1,000 billion times more numerous than stars. If so, there could be several of them in a volume comparable to that of our solar system. In Chapline's view, it would be well worth looking for these mini-holes; if they could be used as a source of energy, they would obviously be of great importance to the economy of our planet. The science fiction world of a town built around a black hole is constructed with such seriousness that Chapline himself loses sight of the most important point: the presence of a mini-hole in the solar system would, first of all, be important as a proof that this whole category of objects really exists, and then as a source of energy.

A couple of years earlier, two scientists from the University of Texas, J. D. Jackson and M. P. Ryan, had suggested that the 1908 Siberian bolide might have been a mini-hole. As you may recall from chapter 1, the fall of that object remained a mystery for quite a long time, mostly because of the very few material traces it left on the ground. The "immaterial" character of a black hole would fully satisfy this requirement. After hitting the ground at a 30° angle with the horizon, it would have gone through the earth, and come out in the North Atlantic at 40° to 50° latitude north, and 30° to 40° longitude west. There was only one way to confirm this hypothesis: find evidence of the body's exit. In open sea and without eyewitnesses, the only evidence could come from the microbarograms of the time, which should have recorded a second atmospheric shock wave similar to that observed during entry. No evidence was found. Furthermore, several scientists refuted Jackson's and Ryan's hypothesis principally on the basis of the material left on the ground (scant, but real), which tends instead to support the cometary theory we have already discussed.

Thus we are forced to conclude that, even if there are mini-holes in the solar system, at least that time they left us alone.

WHITE HOLES
When we started our discussion of black holes, we began with a star whose outer layers, upon the sudden disruption of equilibrium, commence falling inward with ever-increasing acceleration. This collapse eventually causes a singularity in our space-time, which pulls matter away from the surrounding universe and swallows it without remission. Then we attempted to find some of these monsters predicted by theory, and in so doing discovered the amazing fact that they might appear in different guises. Up to now, however, we have not considered a different facet of this question, that is, the possibility that there might be phenomena exactly the opposite of black holes. At some point in our space-time, there might be regions where matter is expelled rather than swallowed—objects that do not collapse but explode. We have not thought there might be black holes in reverse.

If we search for this new kind of monster, we can find at least one: our entire universe. The recession of the galaxies, revealed by spectroscopic observations, tells us that the entire universe is expanding. If we now go back in time, starting from the present, we would see all the galaxies rushing toward a single point where 15 billion years ago the entire mass of the universe was concentrated.[10] In that epoch our space and our time came into being, starting from a point. The reverse process of the universe's growth is analogous to the collapse of a star into a black hole. Here, too, we come to a singularity beyond which we cannot go. However, there is a fundamental difference: the singularity of a black hole is the arrival point, while that of the universe is the starting point. In other words, the universe can be considered as an immense black hole in reverse.

In 1964 the Soviet astrophysicist I. D. Novikov formulated a very intriguing theory based on this very concept. According to this theory, a number of "fragments" of the initial singularity did not take part in the Big Bang but were expelled without exploding. The explosion of these fragments occurred later in the form of many miniature big bangs. These singularities, in which the exact opposite process to gravitational collapse takes place, are called white holes. They should exhibit essentially two properties: an expansion of explosive character and the emission of an enormous amount of energy. The search for objects with these properties, Novikov claimed, was not difficult at all. On the contrary, it was their very existence which had led him to the formulation of his theory. In his view the white holes could only be the nuclei of giant elliptical galaxies, Seyfert galaxies, and quasars.[11]

Novikov's theory complements the speculative ideas proposed by V. A. Ambartsumian around 1950. According to the Armenian astrophysicist, a large amount of matter can remain for a long time in a state that is neither stellar nor gaseous, that is, in neither of the two states in which matter is normally observed in the universe. These masses were called "D bodies." Ambartsumian proposed that from the explosion of these bodies stars would be formed, tied together as expanding associations such as that we observe at the center of the Orion nebula.[12] In 1962 Am-

10. See *Beyond the Moon*, pp. 320–321.
11. See *Beyond the Moon*, chapters 4 and 10.
12. See *Beyond the Moon*, pp. 179–180.

bartsumian pointed out that large amounts of matter could be expelled from the center of galaxies, and he maintained that the violent phenomena observed in giant galaxies and in radio galaxies were due to activity in the nuclei. The cause of this activity could not be explained. By placing fragments of the initial singularity with explosive properties at the center of active galaxies, Novikov's theory has brought Ambartsumian's ideas one step further and provided an explanation for the "D bodies" connected with the origin of the universe.

In 1965 an analogous theory was independently proposed by Y. Ne'eman, and in late 1974 these concepts were further discussed by Y. Gribbin. The latter, in particular, called attention to the fact that white holes in giant elliptical galaxies should have masses of the order of one billion suns. You will recall that Wolfe and Burbidge had estimated the same value for the mass. However, there is one fundamental difference: while they assumed that there was a black hole in the nucleus, Novikov and others proposed that the nucleus would be instead a white hole originating from a residual singularity, around which the galaxy is forming and continuously growing.

The ideas we have just discussed have turned the tables on us, by showing that there might be singularities that do not swallow matter—like black holes—but eject it. In other words, there might be holes that are points of exit rather than of entry. Naturally the existence of white holes would not necessarily preclude that of black holes. On the contrary, this new vision, although more complicated, becomes in a sense more familiar.

One of the reasons why black holes make such a deep impression on us is the sense of ineluctability inherent in them. In our world, up to now, only one thing has appeared completely irreversible: death. In reality nobody dies just at the last moment of his life. One dies at every instant, since no past instant can ever be recaptured again. Time is a dimension we can travel in only one direction, and, upon reflection, its passage appears as the only truly irreversible process.

Theory dictates that for the observer inside a black hole time and space are exchanged in such a way that the direction in which proper time increases is that along which the value of the radius (the spatial coordinate) decreases. Just as the observer cannot return to a higher value of the radius by climbing up the funnel of figure 5.1, he cannot step backward in time. A black hole therefore appears to us as the most frightening embodiment of irreversibility.

Now, the fact that in a white hole the opposite might also exist is somewhat comforting, because it suggests that after we go down the one-way street of the black hole we might return through the one-way street of the white hole.

Once we start considering such fantastic ideas, we may find the courage to ask a naive question, which perhaps was on the tip of our tongue from the beginning: But where do these black holes lead?

In the case of a Schwarzchild black hole, we can answer "to a singularity in our space-time," or, in a more practical manner, "nowhere in particular." But as we have learned, there are other types of black holes, including Kerr's rotating black holes. We have also pointed out that, if there are black holes in the universe, almost certainly they must be of this type or of the Kerr-Newmann type, since all known celestial objects— and therefore also the supermassive stars from which black holes originate—are endowed with rotation. In the case of rotating black holes, theorists tell us that, after crossing the event horizon, one does not necessarily fall into the singularity.

Let us now go back to the gravitational collapse and imagine that our star is becoming ever smaller. As the collapse progresses and the mass is concentrated into an increasingly smaller volume, the plane representing our universe sinks deeper and deeper, and the curvature of space becomes increasingly more pronounced (figure 5.1b and c). However, Einstein and Rosen found that past the event horizon the curvature becomes less pronounced and opens up into a second universe as flat as the first one (figure 5.12). Thus space geometry around a black hole appears very simple: half of it connects our universe to the event horizon, and the other half leads to another universe, identical to the first and also asymptotically flat. The geometric configuration we have just described is called a "worm hole," or an "Einstein-Rosen bridge."

By constructing the Einstein-Rosen bridge, we were under the impression that it connected two identical universes. This is not necessarily

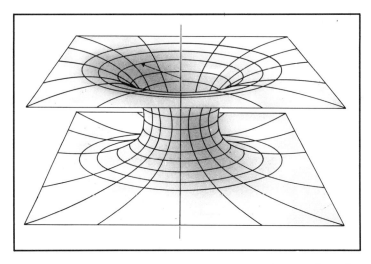

Figure 5.12 An Einstein-Rosen bridge connecting two flat universes. (From C. W. Misner, K. S. Thorne, and J. A. Wheeler, *Gravitation.* Freeman, © 1973.)

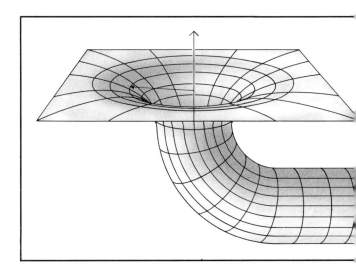

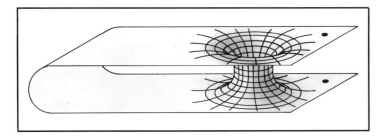

Figure 5.13 An Einstein-Rosen bridge connecting two points in the same universe. (From C. W. Misner, K. S. Thorne, and J. A. Wheeler. *Gravitation.* Freeman, © 1973.)

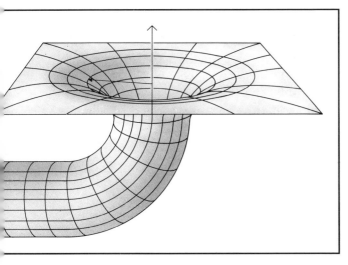

Figure 5.14 Another possible connection between two points in the space-time of the same universe through an Einstein-Rosen bridge. A connection of this type is also called a "wormhole."

true, because it might actually connect two points of the same universe. We only need to imagine a universe shaped as the configuration in figure 5.13, or a hole like the one drawn in figure 5.14. In the hypothesis exemplified by figure 5.12, the black hole would be a passage leading from one universe to another. In the other two cases, it would be a sort of shortcut through which one could rapidly reach points in the universe very remote in space and time by following a relatively short path. Naturally, if our side of the Einstein-Rosen bridge was a black hole, the other side—opening into another universe or into another point in the space-time of our universe—should be a white hole, an opening from which matter and energy swallowed by the black hole at the other end continuously stream out.

Here then we have another type of white hole. The nuclei of peculiar galaxies and Seyferts, as well as quasars, could now be interpreted as exit points of matter coming from another universe. By an analogous mechanism, these same objects could also be explained as white holes at the endpoints of worm holes which suck in matter from other regions of space-time of our universe where, at the entry points, they would appear as black holes. In this hypothesis it is possible for a Seyfert galaxy 100 million light-years away to be ejecting at this moment gases sucked in from a different part of the universe 10 billion years ago. It is even possible that a quasar 10 billion light-years away was formed 10 billion years ago from matter coming from a future time, that is, from a black hole which, as far as we are concerned, has just been born.

JOURNEYS THROUGH SPACE-TIME

The extraordinary concept that by entering a black hole one can exit into another universe can be investigated more thoroughly by using the solution of the field equations found by R. P. Kerr in 1963. Kerr's mathematical deductions can be better understood with the help of the so-called Penrose-Carter diagram. Before we draw it, we should examine the graph in figure 5.15, with which we can express our journeys through space-time.

Space is plotted on the horizontal axis (abscissa), and time on the vertical axis (ordinate); the only condition we impose is that the luminous

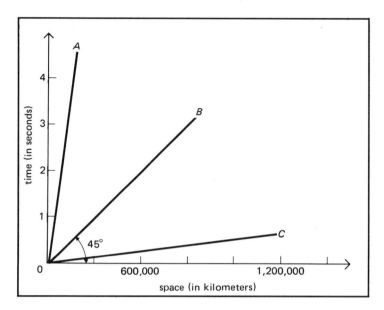

Figure 5.15   Journeys through space-time. Trip 0–*A* is possible because it oc-curs at a speed smaller than that of light. Trip 0–*C* is not possible, since it exceeds the limiting case 0–*B*, corresponding to a trip at exactly the speed of light. (From *Mercury*.)

rays should travel on a line at a 45° angle. This is easily done by choosing appropriate units, for instance, seconds as the unit of time and the distance traveled by light in one second as a unit of space (which is about 300,000 km). Inspection of the graph shows we can move in certain directions but not in others. For example, assuming we measure space and time from a starting point and moment, we find that we can certainly go from 0 to A, but, if we wish to go from 0 to B, we have to travel 300,000 km/s, which only light can do. The journey from 0 to C is outright impossible, since we would have to travel a great distance in a very short time, shorter than the time light would take. As we know from the restricted theory of relativity, the speed of light cannot be exceeded. Thus the possible journeys through space-time, as represented by the graph in figure 5.15, are only those comprised between the ordinate and the line at a 45° angle. To move along the vertical axis means to let time run while remaining stationary; along the line at 45° it means to travel at the speed of light; all the intermediate lines correspond to journeys at the various permissible speeds.

Let us turn now to figure 5.16 which represents Kerr's solution interpreted by the Penrose-Carter diagram. At first sight it may look very complicated, but it is not too difficult to extract from it the information of greatest interest to us. First of all, we must mention that in his solution of the rotating black hole Kerr found not one but two event horizons. The two horizons are represented by the two perpendicular lines that intersect each other at the center of the figure and by all other lines parallel to them. The diagram shows that when all these lines intersect one another, three different kinds of regions are formed. Regions of type I are those between the first event horizon and the external universes, our own or others. Regions of type II are those between the first and second event horizon. Finally, regions of type III are the areas between the second event horizon and the singularity, which in the figure is shown as a wavy line.

The first striking feature in this diagram is the existence of a variety of different universes. The figure shows only four: our own, and three others. But the diagram can be repeated indefinitely in the vertical direction so as to form an infinitely long strip on which the same schematic

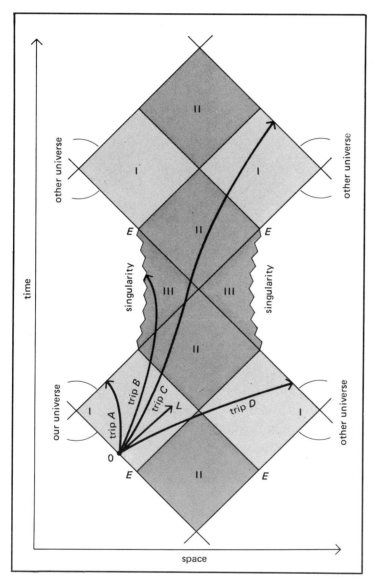

Figure 5.16   A Penrose-Carter diagram illustrating journeys through other universes. (From *Mercury*.)

representation can be continuously repeated. Thus the rotating black hole can be seen as a bridge connecting an infinite number of universes which would otherwise be separate.

The second interesting feature is the singularity, which appears 2-dimensional and parallel to the time axis.

Let us now follow an astronaut willing to embark on this perilous journey toward the unknown with little hope of coming back. Although brave and willing to attempt an impossible task, he cannot undertake journeys at speeds higher than the speed of light. Thus, starting from 0, he cannot reach another universe with a trip of type D. However, he can set off in any direction comprised by the time axis and the $0-L$ line, which subtend an angle of 45°. Within these limitations, he can attempt to explore a rotating black hole. He decides to start off on route A. This route, however, does not take him anywhere near the black hole, and, after leaving earth, he only reaches some other point in our own universe. The astronaut then tries again along route B, which takes him directly to the black hole. He goes through the first event horizon, then through the second, but this time he has made the fatal mistake of aiming directly for the singularity. Inevitably, he will be crushed to death. This tragic end could have been avoided, since one of the most startling properties of Kerr's solution is the fact that one can cross the event horizon without being forced to bump into the singularity. Another astronaut, more cautious and skillful, will choose route C and, with appropriate adjustments of the rocket's thrust during the course of the trip, will have a successful flight. Leaving behind our universe, he first crosses the external event horizon, penetrates into the region between the two horizons, then crosses the internal horizon and comes closest to the singularity without falling into it. Now he crosses the inner event horizon a second time, at a different point, and again passes through the region between horizons. Finally, he crosses the external horizon for the second and last time, leaves the black hole, and emerges into another universe.

At this point the journey is over; but it could go on, taking the astronaut into a third, or a fourth universe, and so on, along the sequence of universes obtained by continuing the diagram in figure 5.16. It is not clear that it could go on ad infinitum, however, since the strip might not be open, as it appears on the page of this book, but closed as it would be if we glued the page on a cylinder (figure 5.17). In this case the astronaut would eventually come back to our own universe, emerging in a certain point of space-time. He might reappear a very short time after leaving earth in some place very far away from our planet; or, he might return exactly to the starting point, but in another time, say, 2 million years after the start of his journey, or 10,000 years earlier. The astronaut could actually choose the time and place in our universe to which he wishes to return simply by setting the flight's parameters in the appropriate manner. We could even go one step further: we could send an unmanned probe to explore all the possible universes connected to us through the black hole and regulate its journey so that after traveling for millions of years, it comes back the next day. The information brought back by this probe would be infinitely richer and more varied than anything—real or imaginary—that man has experienced since the beginning of time.

HOWEVER . . .

Our journey through the black holes is now nearing the end. We have ventured into their world, predicted their origin, and descended into their interior to study their structure and processes. We have found that they are immense reservoirs of energy and that one day perhaps they will open the door to journeys in time. We have seen them in their ambivalence—monstrous and fascinating, graveyards for every living thing and windows opening onto alien skies. But we still do not have positive proof of their existence.

On the other hand, their study has started and developed in a unique way. Contrary to most of the known celestial objects and phenomena, black holes have not been *discovered*, but in a way have been *invented*. If, by discovery, we mean the gathering of a number of observational data all pointing to one explanation, we must admit that this is not what happened for black holes, which are instead the embodiment of a purely mathematical solution to an astrophysical theory. This process has all the earmarks of invention which, all told, is nothing but the construction of a mechanism based on a logical process of one's mind.

This being the case, we can look for proof of the existence of black holes in two ways: either by demonstrating the mathematical theory from which they originate actually corresponds to the observed physical world, or through the discovery of phenomena predicted by theory as being associated with black holes, which cannot be explained in any other way.

The first point is the more sensitive because it seems to cast doubt on the theory of general relativity, which today is widely accepted. In effect, for some time now gravitational theories have been formulated which, at least in the higher-order predictions, do not always agree with general relativity. Among recent theories slightly different from the theory of general relativity is one formulated by H. Yilmaz in 1973. According to Yilmaz, the singularity that appears in Einstein's solution of the field equations disappears, thereby negating the formation and the very existence of black holes. Along the same lines is the theory proposed by D. H. Menzel in 1975. His conclusion was that Schwarzschild's solution is only one particular case of an infinite number of solutions that generally do not correspond to black holes. In his opinion the universe would harbor supermassive objects with properties similar to those observed in quasars and Seyferts, which we will discuss in the next chapter.

Apart from these doubts, even assuming that theory fits the physical world so well as to explain it with the greatest accuracy, it is not clear that all the predicted consequences will actually occur. For instance, even assuming that the theory of general relativity is fully applicable, and that the theoretical deductions derived from it are completely correct, black holes might not form simply because stars with masses larger than 3 $M_\odot$ could somehow shed mass and become smaller than 3 $M_\odot$ before reaching the phase of collapse. As early as 1935 the great theorist A. Eddington, having conceived of something similar to a black hole as the end stage of massive stars and being unprepared to accept the absurdity of such a solution, wrote, "I was forced to conclude that this was almost a

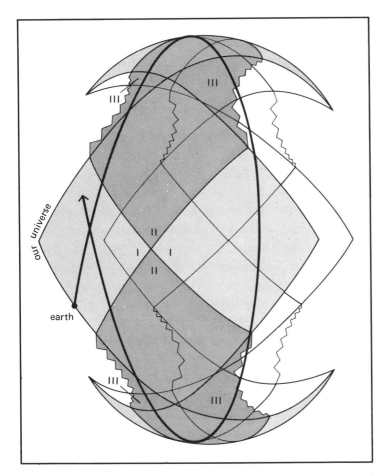

Figure 5.17   A Penrose-Carter diagram for a closed ensemble of universes is illustrated by wrapping a planar Penrose-Carter diagram around a cylinder. (From *Mercury*.)

reductio ad absurdum of the formula of relativistic degeneration. Several accidents can intervene to save the star, but I ask for a greater protection. I think that there should be a law of nature that prevents the star from behaving in such an absurd way."

The second way of proving the existence of black holes is more promising. The discovery of an ever-increasing number of phenomena that cannot be explained in any other way would eventually grip the black hole in a vise of evidence which we could reasonably accept as conclusive proof of its existence. However, even in this approach doubts have been raised with regard to the compelling nature of the available evidence. J. N. Bahcall and J. Pringle have proposed models to explain current observations without invoking black holes. For instance, in the specific case of Cyg X-1, one can make the case that the invisible companion is not a black hole but another normal star of about 6 $\mathcal{M}_\odot$ with a satellite neutron star emitting X rays revolving about it. Alternatively, one could think of a neutron star revolving in a wide orbit around a perfectly normal binary system. These models, although less plausible, are possible.

Furthermore, in the spring of 1975, Avni and Bahcall published a study that further weakened the evidence for black holes based on X-ray binaries. According to these authors, the conclusion that the invisible component in Cyg X-1 cannot be a normal star has been reached because the available light curves of the primary star had been interpreted according to a standard model which may not correspond to reality in all cases.

For instance, it had been assumed that the primary star's rotation was synchronous with orbital motion; that the star was in hydrostatic equilibrium in the co-rotating region, so that von Zeipel's theorem could be applied; and, finally, that it would fill its Roche lobe.

Avni and Bahcall undertook the study of four X-ray binaries, taking into account the theoretical, parametric, and observational uncertainties that could invalidate the standard model. They also considered other effects such as the mutual heating and reflection effects of the two stars. Their goal was not to determine the best values of the masses starting from a fixed scheme—which after all could be too restrictive, or even inconsistent, with the physical reality of the system—but to determine all possible values of the masses of the two components and their sensitivity to variations in the initial model.

Following this approach, they found that Cyg X-1 is not necessarily a black hole. In fact, using this more general approach, the observational data on this system can be satisfied by 18 solutions in which the primary star has a mass ranging from 23 to 36 $M_\odot$ and a radius about 20 times that of the sun. In 6 of the 18 solutions the secondary star is a black hole with a mass from 8.4 to 11.2 $M_\odot$; in 9 cases it is a normal star of the earlier spectral types, still with a mass between 8.4 and 11.2 $M_\odot$ and a radius of 5 to 7 times that of the sun; in the last 3 solutions the secondary is a star above the main sequence or a peculiar one. Thus out of 18 possible solutions only 6 correspond to a black hole.

According to Avni and Bahcall, on the basis of the observations published to date, it would be very hard to find a companion with the properties they have assigned to it. However, they estimate that this could be done with spectrophotometric observations with a precision of 1 percent, since they could select among the 18 the solution that best satisfies the observations and that could be accepted as the correct one. Avni and Bahcall then do not just take apart a result perhaps too readily arrived at but also propose a new way of reaching the conclusion closest to reality.

It could still turn out to be a black hole; but until that day, we will not know with any certainty whether we have glimpsed a new and mysterious facet of reality which will add a new dimension to our knowledge of the universe or whether we have read a fascinating chapter of science fiction.

This conclusion is not disappointing. On the contrary, it is very exciting because the quest is still going on. It could be in our generation or in the next that someone searching for this reality opens a new window on the universe. Then we will know if it really is an opening through which we can glimpse another world, or an opaque screen where the imaginative minds of scientists have projected an unreal landscape, beyond which there is nothing to be found.

6 THE KEY MONSTERS

BL LACERTAE OBJECTS

A new variable star was discovered in 1929 at the Sonneberg Observatory—one of the many discovered at that institute which specialized
in the search and study of these stars. From the onset the star showed
irregular light variations with an amplitude of about 3 magnitudes
but no other peculiarity. Since irregular fluctuations in brightness
may be due to a number of different reasons, variables of this type
constitute a very heterogeneous group. Owing to their irregular behavior—which significantly reduces the chances of interpretation—
they tend to be ignored by astronomers, and consequently the new variable was almost forgotten. A monumental work published in 1957,
Geschichte und Literatur des Lichtwechsels der veränderlichen Sterne, contains
the history of every known variable star, as well as a list of all the works
written about them. While several pages were devoted to some of these
stars, all that could be said about BL Lacertae was covered in one line
(see vol. 4).

Interest in this star was suddenly kindled in 1968. Two Canadian astronomers, J. M. Mac Leod and B. H. Andrew, found that one of the
radio sources just discovered at the University of Illinois had a peculiar
radio spectrum, and J. L. Schmitt found that this new radio source coincided with BL Lacertae. In little more than a year many new and interesting discoveries followed. It was found that radio emission from the
star appeared to vary, often in as short a period as one month, that its
radiation was polarized, and that also the polarization varied, sometimes
from one week to the next. In addition, it was discovered that the star
could undergo rapid irregular fluctuations also in visible light, changing
even from one night to the next. This kind of behavior is already pretty
strange in itself, but the star's most astonishing feature emerged from
observations of its color and spectrum. In late 1968 BL Lac was observed
with the 5-m telescope at Mount Palomar and with the 2.10-m telescope
at Kitt Peak; its optical spectrum turned out to be completely featureless.

This particular characteristic has been found almost exclusively in
some white dwarfs, but precise measurements of the star's U-B and B-V
color indexes showed that it could not be a white dwarf. When the star

was placed on the color-color diagram on the basis of the values found
for the two indexes, it did not fit in with white dwarfs but fell instead
between quasars and N galaxies (figure 6.1).

We will discuss shortly these two classes of objects. But it is already
quite obvious that such a result was more than sufficient to suggest
the possibility, never considered until then, that BL Lac might not be a
star at all. Rather, it appeared to be an extragalactic object of a peculiar
type that did not correspond to any of those known: it was too red to be
a quasar and too blue to be an N galaxy. On the other hand, its apparent
magnitude was relatively high, and, if it was roughly at the same distance
as the nearest quasars, it would have been the brightest object known in
the universe. Unfortunately, the all-important question of the distance
could not be resolved. Since the star's spectrum was totally devoid of
lines, it was impossible to say if it was shifted toward the red—let alone
measure the distance from the extent of the shift, following the method
commonly used for galaxies and all other extragalactic objects.

Apart from these uncertainties, while photographing BL Lac with
Mount Palomar's 5-m telescope, H. Arp noticed that the star's image was
not exactly stellar and suggested instead that it might be a compact
galaxy or a very peculiar planetary nebula. The first round of research
came to a close in mid-1969 with an exhortation to all astronomers to
continue the study of "this highly unusual, perhaps unique, object."

But BL Lacertae was not unique. Concurrently with the discoveries we
have just mentioned, it was found that another radio source exhibited
similar properties. It had been independently discovered in 1968 by
radio astronomers from three different observatories and designated OJ
287. A short time later the peculiarity of its radio spectrum was pointed
out by the same astronomer, B. H. Andrew, who had first noticed it in
BL Lacertae. In 1970 G. M. Blake discovered that OJ 287 was also a visible object of stellar appearance, and in the spring of 1971 observations
made by T. D. Kinman and E. K. Conklin at Kitt Peak Observatory revealed that this object exhibited variations in brightness as rapid as a few
days. Like BL Lacertae it was found to have a continuous spectrum featureless in the blue and a highly variable polarization. That same year
G. W. Brandie and M. A. Stull, while observing OJ 287 with the 26-m

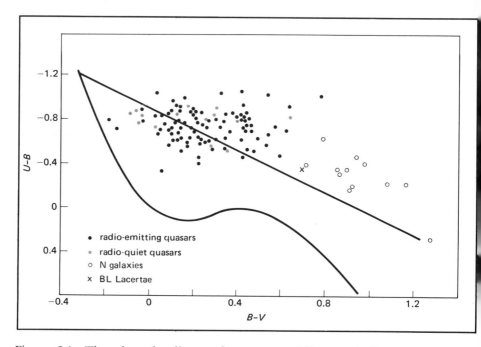

Figure 6.1   The color-color diagram for quasars and N type galaxies. BL
Lacertae falls between the two groups. The straight line corresponds to black
body emission, and the curved one to unreddened stars of the main sequence
(Hyades cluster). They are used as references. (Adapted from A. Sandage,
Sémaine d' étude sur les noyaux des galaxies.)

radio telescope at the University of Michigan, discovered that its radio emission was also variable.

The similarity between OJ 287 and BL Lacertae was now quite evident. Then, when three more objects were found to exhibit similar properties, it became obvious that the study of a peculiar and seemingly unique case had led to the discovery of a new type of celestial objects, as interesting as they were puzzling. The three new objects were ON 325, ON 321, and PKS 1514-24. The last two were already known to optical astronomers as variable stars and had been designated W Comae and AP Librae. All five objects were discussed in a paper published in July 1972. In it P. A. Strittmatter and five other astronomers defined for the first time the basic properties common to stars of this new class.

It was now clear that the new class comprised objects of stellar appearance (figure 6.2), nearly pointlike even in radio emission, which exhibited the following characteristics:

1. rapid variations in the intensity of the radiation emitted at radio, infrared, and visible wavelengths;
2. a distribution of energy in the spectrum increasing from visible to infrared, and thus showing maximum emission in the infrared;
3. lack of any features (either lines or bands) in the low dispersion optical spectra;
4. intense and variable polarization at both visible and radio wavelengths.

Although the new class was now well defined, there remained the problem of explaining what such objects could be and whether they were near or far, galactic, or extragalactic.[1]

However, on the basis of the observed features astronomers could now search for all objects of this type yet to be discovered. By obtaining a greater number of examples, they could find out how rare these objects

1. The new objects were called "Lacertids." Although practical, this choice is less than perfect because the term can be confused with those used for meteor showers (Leonids, Perseids, and so on). Currently they are called either "Lacertids" or "BL Lacertae objects."

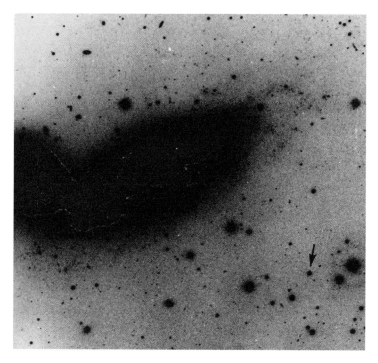

Figure 6.2   A BL Lac object (indicated by the arrow) photographed in the same field as the galaxy NGC 6503. Note its stellar appearance. (Courtesy of H. Arp.)

were. Moreover, they could hope to discover some that, either because of their greater brightness or because of additional peculiarities, would be easier to study and even suggest a similarity with known objects.

The hunt was soon under way, using a method suggested by the events leading to the discovery of the first Lacertids. Since almost all of these objects had already been known for years as variable stars, the *General Catalog of Variable Stars* might easily hide in its pages a great number of Lacertids under the misleading, but innocent-looking, label of "irregular" variables. A comprehensive list of all the radio sources discovered to date published in 1970 by R. S. Dixon greatly facilitated the search through the catalogue, enabling astronomers to compare the positions of the radio sources with those of the suspected variables. This selection produced about a hundred candidates; unfortunately, further examination revealed that none of them was a BL Lac object.

In the meantime other astronomers had been using a different approach. The radio spectrum of these objects is rather flat, and they attempted to find optical counterparts of the radio sources exhibiting this property. This approach found increasing favor and proved to be quite fruitful; the number of known or suspected Lacertids reached thirty by mid-1976, Thus the material available today is much more abundant than it was at the beginning of the search, and at last some reasonable explanations for the phenomenon are being proposed.

Actually interpretations had not been lacking from the very start. Strittmatter and his collaborators had suggested that the Lacertids might be objects with large blue shifts. This theory had already been advanced in the case of quasars and was highly controversial.

When the large red shifts of the quasars were discovered, not all astronomers concurred with the theory that they were of a cosmological nature, and even today this difference of opinion has not been completely resolved. Those who did not accept the cosmological interpretation suggested that the red shifts might be due to a real motion of the quasars with respect to the galaxies. Assuming that an extragalactic object, for instance, a galaxy, had exploded in the past and violently ejected some fragments, the spectra of all the fragments moving away from the explosion, and also away from us, should show red shifts proportional to their radial velocities. Of course the explosion would have scattered fragments in all directions. Those coming toward us should exhibit a blue shift corresponding to a velocity approximately equal to that of the fragments which are receding. In other words, in addition to quasars with red-shifted spectra, there should be others with blue-shifted spectra. Quasars of this type had never been observed, however, and this cast doubts on the whole theory.

On the other hand, in 1967 the Burbidges had calculated and discussed the optical properties that should be exhibited by blue-shifted quasars. It now appeared that many of the predicted properties were actually present in the Lacertids. Thus it was suggested that these objects represented the long-sought proof of the noncosmological theory of quasars. In this case the Lacertids would still be extragalactic but nearer, smaller, and less luminous than they would be if they were located at the same distance as the quasars, and if we accept the cosmological interpretation of quasars.

However, not everybody agreed with the view that the Lacertids were extragalactic. In July 1974 S. T. Shapiro and J. L. Elliot proposed a completely different theory. Elaborating on a suggestion made a year earlier by J. E. Pringle and his collaborators, the two scientists from Cornell University showed that BL Lac and similar objects could be isolated black holes accreting interstellar gas. With this model they could explain the visible and infrared spectrum, the rapid variations in intensity and in polarization, and the lack of features in the optical spectrum. This possible interpretation made further study of the Lacertids even more important. If it could be confirmed that these objects were galactic, there would be new candidates for the black holes; conversely, if they turned out to be extragalactic, Shapiro and Elliot's theory would no longer be applicable. Thus now more than ever it was absolutely essential to determine whether Lacertids were extragalactic or not.

J. B. Oke and J. E. Gunn undertook this research starting from a very simple premise. They accepted the serious drawback that the spectrum of BL Lac was featureless and therefore could not provide a basis for discovering and measuring the red shift. On the other hand, they also considered the possibility that BL Lac might not be the *whole* object but only *part* of a larger and fainter one; possibly the bright stellar nucleus of a galaxy similar in appearance to a Seyfert.[2] Starting from this hypothesis, in October 1973 they set out to look for this possible galaxy. They built a ring diaphragm that would allow them to cover BL Lac, which was much too bright, and to observe only the light coming from the surrounding region (corresponding to the ring) where the much fainter galaxy was supposed to extend. Using this device on a spectrometer mounted on the 5-m telescope at Mount Palomar, they were able to obtain a different spectrum where the H and K lines of calcium, the G band, the H_β line of hydrogen, and a line of neutral Mg were just barely discernible. These are normally the most conspicuous features we observe in the spectra of giant elliptical galaxies. Hence, surrounding BL Lac, and corresponding to the ring, there was indeed a galaxy of which our object was the nucleus. Moreover, the energy distribution in the spectrum, which was determined at the same time, appeared to confirm that it was a giant elliptical galaxy. But this was not all: the observed spectral features were not at rest, as we would see them in our laboratories, but showed a red shift of $z = 0.07$. From this value it could be determined that BL Lac and the galaxy associated with it were receding at a speed of 21,000 km/s and were at a distance of 380 Mpc, or 1,240,000,000 light-years. At this great distance BL Lac, with an apparent magnitude of 13, had to have an absolute visual magnitude $M_v = -22.9$. Thus it appeared to be far more luminous than our whole galaxy and undoubtedly one of the brightest galaxies we know.

This result was published in 1974 and appeared uncontestable. But less than a year later J. A. Baldwin et al. published observations that were in disagreement with those of Oke and Gunn. Observing BL Lac with the 3-m reflector at Lick Observatory, they found no evidence whatsoever of the elliptical galaxy. Consequently, they denied its existence and suggested some technical reasons why Oke and Gunn could be mistaken. A few months later Kinman cast doubts on the latter finding by remarking in one of his works that, according to observations he had made with

2. For a description of the main characteristics of Seyfert galaxies, see *Beyond the Moon*, pp. 274–280. An in-depth examination of these galaxies will be found later in this chapter.

the 2.10-m reflector of Kitt Peak, the distribution of surface brightness was consistent with that of a giant elliptical galaxy of $z = 0.07$.

This disagreement was conclusively resolved by Oke and Gunn in late 1974. In collaboration with T. X. Thuan, they observed BL Lac again with Mount Palomar's reflector. But this time they made efforts to place themselves in the best possible conditions by observing only on nights of excellent visibility and at the times when BL Lac was at the zenith and its brightness was low, so as to minimize the interference of the bright nucleus in the observation of the surrounding galaxy. As a result of these precautions, the previously observed lines appeared again—quite conspicuously. In late 1975 the authors published their results, which confirmed that BL Lac was undoubtedly at the center of a giant elliptical galaxy exhibiting a red shift of $z = 0.07$.

In the meanwhile, lines were found and measured in the spectra of two other Lacertids: AP Lib and PKS 0735+178. The former was found to have a red shift $z = 0.049$, equivalent to a distance of 960 million light-years; in the latter z was measured at 0.424, corresponding to about 8 billion light-years.[3]

Thus in early 1976 there were three Lacertids with a measured red shift. Assuming that the red shifts were cosmological, that is, due to the general expansion of the universe, it could be concluded that at least three of these objects were extragalactic, very far away, and extremely luminous. The extragalactic nature of BL Lac objects and their association with galaxies were thus definitely proved.

At the same time other astronomers were gathering additional information. Following Oke and Gunn's first communication of 1974, the possibility that Lacertids might be associated with galaxies found increasing acceptance and led to the investigation of those galaxies that like the elliptical ones might harbor these objects.

Following this approach a number of interesting facts came to light. The results of a study of fifty-seven elliptical galaxies identified as radio sources were published in the summer of 1975. Three of these galaxies,

3. For the determination of extragalactic distances in light years through radial velocities or the values of the red shift z, see the appendix.

NGC 6454, B2 1652+39, and B2 1101+38, exhibited flat radio spectra at high frequencies similar to those of the Lacertids. Spectrographic, photometric, polarimetric, and radio analysis not only confirmed that these three galaxies were elliptical but also revealed additional features which emphasized their likeness to BL Lac, AP Lib, and 0735+178, notably, the presence of a compact nonthermal source at the center similar to the Lacertid observed in the core of giant elliptical galaxies. The newly discovered Lacertids were fainter than the known ones and exhibited optical properties that placed them between normal elliptical galaxies and BL Lac objects.

All the research in this field has thus confirmed the connection between Lacertids and galaxies. In addition, it has been found that this connection may not be limited to elliptical galaxies. Certain BL Lac objects appear to be related to Markarian, Seyfert, and N galaxies, and perhaps even to quasars. We will discuss these objects shortly in greater detail. Right now we only wish to point out one common characteristic: all these extragalactic objects are so compact and luminous that they look like stars, for which in fact they are often mistaken. This is true also for BL Lac objects. Furthermore, they not only appear to be small but are intrinsically so, at least when compared with the dimensions of normal galaxies. Normal galaxies are immense systems with diameters of tens or hundreds of thousands of light-years, and even their nuclei, although relatively small, have diameters of the order of hundreds or thousands of light-years. This cannot be the case for BL Lac objects. As we have seen, they are variable, and their fluctuations in brightness occur sometimes in the course of a few days or hours.

The fact that rapid fluctuations imply that light is emitted from a relatively small region is easily demonstrated. Let us assume (figure 6.3) that the phenomenon occurs at the same time over the entire surface of the object. If the object had a diameter of 1,000 light-years, the ray of light originating from A should arrive 500 years later than the ray that left B at the same time. On the other hand, if the object had a diameter of only 1 light-day, the time lag between the two rays would be 12 hours; and taking into account also the rays that start from intermediate points and arrive during the 12-hour interval, the variation would be observed as a single event. Therefore, from the rapidity with which the brightness of the object varies, we can determine its maximum diameter. For BL Lacertae it has been computed at 7 light-days, and for MA 0829+047 at only 22.8 light-hours. In other words, the diameter of the latter appears to be 170 AU, or little more than twice that of the solar system. It is a very large diameter as compared to the distance between the earth and the moon, which is the space man has learned to bridge, but it is nothing when compared to the diameters of galaxies, or even of their nuclei. These dimensions would be more suitable for a fragment of a galaxy. The fact is that this fragment is intensely luminous and emits far more light than that radiated by the billions of stars comprised in a galaxy. We do not know yet why this is so and what processes are involved.

ACTIVE GALAXIES

The optical and radio variability observed in BL Lac objects is an indication that active processes are taking place in them. Evidence of activity has also been found in other extragalactic objects, typically those with starlike nuclei such as quasars and certain peculiar galaxies. Frequently, the broadening of their spectral lines reveals a remarkable turbulence in the central regions and an explosive ejection of gas, which is similar to that occurring in novae and supernovae but far more impressive in view of the dimensions and masses involved. Thus, in addition to the normal galaxies of classic appearance which do not exhibit rapid variations, there exists also a group of strange active galaxies. Their number is not too large (at least as far as we know today), yet their study may prove extremely rewarding in furthering our understanding of extragalactic objects and of the structure and evolution of the universe.

We know only a few thousand active galaxies, but there may be many others that have been mistaken for stars because of their stellar appearance. It has not been possible as yet to classify them with any precision nor to produce a comprehensive interpretation. Up to now astronomers have simply attempted to discover these objects and assign them to different groups on the basis of their distinguishing characteristics. On the other hand, it is only fair to point out that this field of research is very

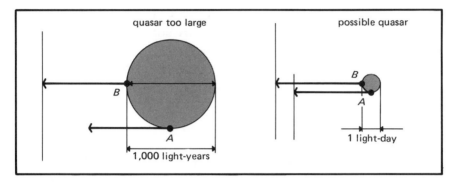

Figure 6.3   The time-lag of light emitted from different regions of a quasar as a function of its size. The observer on earth is assumed to be at the far left.

young. Most of the results we are about to present have come from research that began in 1970. Moreover, they are not definitive results but only some of the most significant findings of recent years. We have gathered them together in some order so that the reader may become acquainted with this field and participate in the search for a theoretical interpretation.

MARKARIAN GALAXIES

The color and spectrum of a normal galaxy shows that the light originating from its central region is very similar to that of the sun. More accurately, the center of a normal galaxy has been found to consist mainly of stars of spectral types G and K. In the early 1960s, however, the Armenian astronomer B. E. Markarian discovered that the central region of some of the most luminous galaxies radiated essentially blue light and appeared to be rich in A and F stars. Further study of these galaxies revealed that the anomaly observed in the spectrum and color of their nuclei was due to an excess in the emission of nonthermal ultraviolet radiation (UV). In other words, the emitted energy was not due to the heat of high temperature stars but to different and unknown processes. The emission appeared to originate from a relatively small region at the very center of each galaxy. This could be demonstrated by placing an occulting disk on the nucleus of the galaxy; the smaller the disk, the greater the excess UV emission appeared to be.

This result was so interesting that in 1965 Markarian initiated a systematic search for galaxies of this type, using the 1-m Schmidt telescope at Byurakan Observatory with a low dispersion objective prism. The goal of this survey was to discover all the objects of this type as faint as magnitude 17 observable from Armenia, in other words, north of Dec. $-5°$ and with galactic latitude $b \geq 30°$. At this writing about half of the program has been completed; 6,000 square degrees of the sky have been explored, resulting in the discovery of 507 objects with excess UV emission. Of these, 10 appear to be double; hence the total number amounts to 517. Today they are known as "Markarian galaxies."

While the survey was under way, Markarian himself as well as other astronomers in the world's foremost observatories conducted detailed investigations of several of his galaxies. Consequently, even though only the first phase of the research has been completed, we have now available a fair amount of interesting data.

First of all it was found that there are two types of Markarian galaxies, designated s and d. Type s consists of galaxies with stellar appearance, and type d of those which appear diffuse. In addition, there are some galaxies of intermediate type: type sd (intermediate but more stellar) and type ds (intermediate but more diffuse). Of the 507 Markarian galaxies found so far, 249 are of type s or sd, and 268 of type d or ds. This appears to be a fundamental distinction that almost certainly corresponds to a basic difference between the two classes of objects. We will now examine them separately.

When observed directly, Markarian galaxies of type d and ds appear rather extended and diffuse (figure 6.4a). Their spectra show a fairly weak continuum but with the ultraviolet excess characteristic of all Markarian galaxies. Recent studies have shown that most of them are similar to compact stellar associations and super-associations of high temperature giant and supergiant stars embedded in gaseous nebulae, such as those that have been occasionally observed in the arms of open spiral galaxies and in certain irregular galaxies. In 1973 D. W. Weedman published a great number of observations in the UBV system,[4] which showed that d and ds Markarian galaxies fall in the region of the color-color diagram that, according to G. de Vaucouleurs, is occupied by blue irregular Magellanic galaxies (figure 6.5). Weedman had calculated in 1972 that the energy emitted by these galaxies in the H_β line of hydrogen is between 2×10^{39} and 4×10^{41} erg/s. The well-known Tarantula nebula in the Large Magellanic Cloud, which is the brightest H II region in the galaxies of the local group, emits in the H_β line a flux of 3×10^{39} erg/s, which falls in the range of values found by Weedman for d and ds Markarians. Thus, adopting these values for the H_β flux, and assuming that the exciting stars are of spectral class O7 with temperatures of about

4. The UBV system gives photoelectrically measured magnitudes corresponding to effective wavelengths of 0.365, 0.44, and 0.55 μ. They are not very different from the old ultraviolet, blue, and visual magnitudes.

40,000°K, we find that the number of these stars in each d and ds Markarian galaxy ranges from 1,000 to 100,000. As a basis for comparison, we should recall that our own Galaxy, although a great deal larger, is believed to contain no more than 6,000 stars of spectral type O. These results suggest the possibility that d and ds Markarian galaxies may be conglomerates of gas and newly formed giant blue stars and therefore that they too may be very young, or, more accurately, that they were very young at the time they radiated the light we presently observe.

Moreover, it has been pointed out that these H II regions, rich in young stars and isolated in intergalactic space, are similar to two of Zwicky's galaxies (I Zw 18 and II Zw 40), which were studied in 1970 by Sargent and Searle and still remain quite puzzling for several reasons.[5] Among the objects of this type there are also about forty galaxies that G. Haro in 1958 had found to be exceptionally blue from plates in three colors. These objects, called Haro galaxies, have now come to be part of the larger group of d-ds Markarian galaxies; hence there is no longer any reason to segregate them in a separate class.

The second group of Markarian galaxies is far more complex. As a rule galaxies of type s-sd have a very compact nucleus, stellar or quasi-stellar in appearance, around which a relatively faint galaxy can occasionally be detected (figure 6.4b). Their spectra show an intense continuum and an energy distribution similar to that of quasi-stellar objects (QSO). This unsuspected similarity was confirmed by photoelectric measurements which showed that in the color-color diagram many of them fall precisely in the region of the QSO (figure 6.5). Further proof was obtained with slit spectographs, revealing spectra with emission lines, that is, more similar to the spectra of quasars than to those of normal galaxies.

Ninety percent of the Markarian galaxies exhibit emission lines in their spectra. Generally, the emission lines are those of the Balmer series of hydrogen, but lines of other elements, both permitted and forbidden, are also found. This phenomenon, extremely rare in normal galaxies, is of great significance because it tells us the nuclei of these galaxies contain

5. See *Beyond the Moon*, p. 294.

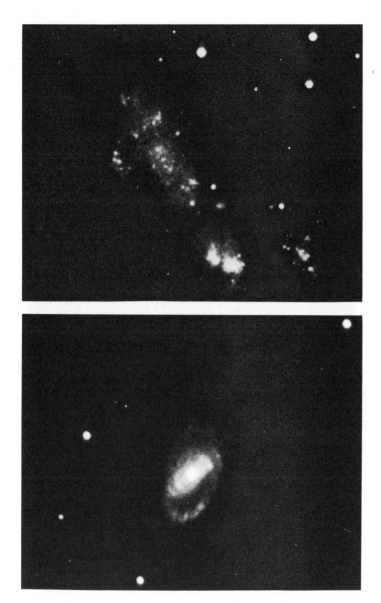

Figure 6.4 *Top*, a Markarian type d object (Ma 71). *Bottom*, a Markarian type s object (Ma 122). Both photographs were obtained by C. Casini with the 182-cm telescope at Cima Ekar (Asiago). (Courtesy of C. Casini.)

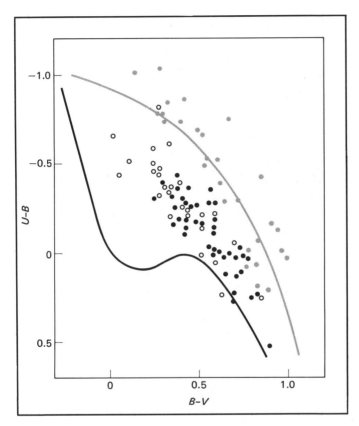

Figure 6.5   The color-color diagram for d-ds Markarian galaxies (open circles), BN galaxies (black dots), and Seyfert galaxies (gray dots). The black reference curve corresponds to unreddened stars of the main sequence; the gray curve defines the locus of positions on the color-color diagram occupied by a normal galaxy, such as M 31 in Andromeda, assuming that its emission was increasingly dominated by the superimposed emission of a typical quasar. (From *The Astrophysical Journal.*)

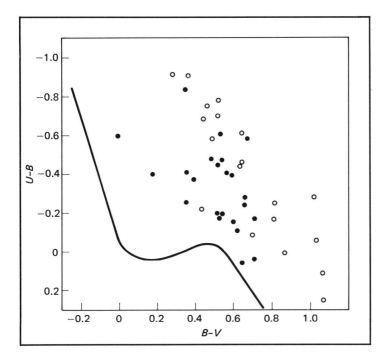

Figure 6.6 The distribution of s-sd Markarian galaxies on the color-color diagram. The open circles indicate objects with broad emission lines; the gray dots those with narrow lines. The black curve represents stars of the main sequence. (From *Astrofizica*.)

large amounts of gas as well as high frequency radiation capable of exciting it. Both the width of the lines and their absolute and relative intensities vary from one galaxy to another. In some of these galaxies the lines are more or less elongated, while in others they are noticeably inclined, revealing an appreciable rotation of the nuclei. Perhaps the most remarkable finding is that 25 percent of s-sd Markarians have broad emission lines, quite similar to the lines that appear in the spectra of those extraordinary objects known as Seyfert galaxies. These strange galaxies have very small nuclei of stellar appearance and variable brightness, which appear to be in a state of permanent explosion. The similarity between these two types of galaxies was also confirmed by the observations carried out during the winter of 1971 to 1972 by T. F. Adams with the 2.30-m telescope at the Steward Observatory. While studying Markarian 78, he discovered that in the inner regions of its nucleus there were radiating gas clouds similar to those observed in some of the Seyferts. Furthermore, by photographing it through interference filters, he discovered that the galaxy was surrounded by a net of filaments like those already observed in exploding galaxies.

As a result astronomers are now convinced that a number of s-sd Markarian galaxies are to be regarded as Seyferts. In fact the number of Seyferts, which for a long time had remained at twelve, has been expanded to fifty-one, assimilating many s-sd Markarian galaxies.

We will return shortly to this subject. But first let us examine the spectra of the remaining s-sd Markarian galaxies. Almost all of them show narrow emission lines like those observed in many QSOs and in some quasars. The difference between narrow and broad lines is explained in this manner: in the first case the gas that produces them is not moving along our line of sight, while in the second it is, revealing an expansion from the center of the cloud outward. Galaxies with narrow lines can be explained by assuming that the radiating gas is either quiescent or moving only in a specific direction, for instance, along the axis of the rotating cloud. In this case, when the axis of rotation happens to be perpendicular to our line of sight, the radial component of the velocity of the ejected gas is nil; hence the line remains narrow. Another possibility is that the gas is not completely quiescent but is ejected from the nucleus at low speed. This explanation would be in agreement with a hypothesis proposed by Markarian. Seyfert galaxies are a very small fraction of all known galaxies; hence the ejection of gas from their nuclei must be a phenomenon of relatively short duration—a few million years at most. Rapid variations in brightness and in the spectral lines reveal rapid changes in the nuclei and suggest that the phenomenon may be discontinuous and recurrent. In this case there would be periods in which a Seyfert is in a quiescent state. At these times the galaxy would retain all the properties of a Seyfert except the broadening of the lines and would therefore appear as an s or sd Markarian with narrow lines.

Radio emission is a very significant facet of the total activity of a celestial object. Consequently, radio observations of Markarian galaxies commenced almost as soon as they were discovered. Until 1974 the results were few and contradictory; then in the spring of 1975 R. A. Sramek and H. M. Tovmassian conducted a radio survey of all Markarian galaxies at a wavelength of 6 cm. Only 28 turned out to be radio emitters; of these, the seven that showed the most intense radio emission were found to be Seyferts. If we recall that there are 51 Markarian galaxies of the Seyfert type, it follows that about 14 percent of all Markarian-Seyfert galaxies are radio emitters. Only 21, or 4.6 percent, of the remaining Markarian galaxies are radio sources.

To summarize, more radio sources are found among Markarian-Seyfert galaxies (which are only 10 percent of all Markarians) than among all the remaining Markarian galaxies.

Mention of percentages brings to mind the fact that we have not yet dealt with a very interesting point, namely, the number of Markarian galaxies as compared to that of all other galaxies. This calculation was often attempted, but the first significant result was obtained by J. Huchra and W. L. W. Sargent when they could avail themselves of four of the five lists published to date by Markarian. By grouping Markarian galaxies on the basis of absolute magnitude, that is, intrinsic luminosity, they found that those with M_{pg} from -14 to -21 represent about 10 percent of all galaxies. For higher luminosities, however, the percentage increases, and for $M_{pg} = -23$ all galaxies turn out to be Markarians, and most of them Seyferts (figure 6.7). This result, published in late 1973, has neither been confirmed nor disproved, but it is evident that further study of this field will add greatly to our understanding of the evolution of the universe. Possibly this will be undertaken when the survey of Markarian galaxies has been extended to cover the whole sky, including the southern hemisphere.

In the last few years a very interesting fact has emerged concerning the age of Markarian galaxies. According to J. Heidmann and A. T. Kalloghlian, these objects are very young. The reasoning is quite simple. Of the 507 Markarian galaxies discovered to date 10 are double, with angular distances smaller than 11 arc minutes (figure 6.8). This rate of occurrence suggests a true physical connection rather than a random superposition along the line of sight. This has been confirmed in four cases, where an examination of the red shifts has shown very similar values for the galaxies of each pair. These systems, however, cannot be stable; calculations show that in order to be stable they should have average masses of at least 1,700,000,000 $\mathcal{M}_\odot$. This value is far greater than that found for every Markarian galaxy for which it has been possible to compute the mass from observations of neutral hydrogen. Hence Heidmann and Kalloghlian concluded that the double Markarian galaxies must have positive energy, which means they must be drawing increasingly farther apart. According to their calculations, the complete separation of the two components—when there will no longer be any physical tie between them—would take place between 150 million and 1 billion years.

C. Casini and J. Heidmann subsequently found that there are also pairs consisting of a Markarian and a normal galaxy (figure 6.8b). In these pairs the distance between the two components ranges from 50 to 120 kpc and thus is generally greater than the average distance of 55 kpc measured in pairs where both galaxies are Markarians. These "mixed pairs" would have existed for a longer time, and the normal galaxy would be an ex-Markarian that has had time to evolve. The age computed for such pairs by Heidmann and Kalloghlian shows that they must have formed long after the beginning of the universe, which, according to the most recent calculations, occurred about 18 billion years ago.

Additional information on the evolution of Markarian galaxies could be obtained from a more refined morphological analysis. Unfortunately, we have very few wide-field photographs capable of revealing their structure. With the exception of the work carried out by Kalloghlian in 1968, the structure of Markarian galaxies can only be seen in the plates of the Mount Palomar sky survey, in which they appear nearly starlike. However, scientists are optimistic that further study of the shape and structure of these galaxies will yield some significant results, since in this case, contrary to what has almost always happened in the past, the initial research was limited almost exclusively to spectroscopic analysis. Casini and Heidmann have recently begun to photograph Markarian galaxies with an appreciable enlargement, using the two largest telescopes of the Asiago and Haute-Provence observatories. Their first set of findings, published in the spring of 1976, seem to point to a new class of galaxies that the two astronomers have called "clumpy irregulars," or Ic, because of their appearance (figure 6.8c). These galaxies would be ten times more massive than classic irregulars and rich in regions where stars are rapidly forming. In time Ic galaxies would be destroyed or transformed into galaxies of more typical appearance.

N AND COMPACT GALAXIES

N galaxies are another important group of extragalactic objects. In 1958 they were defined by W. W. Morgan as systems with small bright nuclei superposed on a significantly weaker background. Six years later Morgan, Matthews, and Schmidt described them more accurately as galaxies with a bright starlike nucleus containing most of the luminosity of the system and a faint nebular shell of limited extension. This definition is practically the same as that given by C. Seyfert in 1943 for the class of galaxies that today bears his name. According to Seyfert, his galaxies are characterized by an exceptionally luminous, stellar or quasi-stellar nucleus containing a relatively high percentage of the total amount of light emitted by the system. The near perfect similarity between the two definitions brought about a certain amount of confusion in the usage of the two terms—N and Seyfert galaxies. Morgan himself had included in his catalogue as N galaxies two typical Seyferts—NGC 4051 and NGC 4151. The main morphological characteristic common to both types of

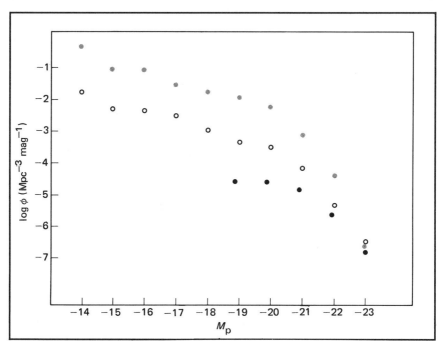

Figure 6.7   The abundance of Markarian galaxies (open circles) and Seyferts (black dots) is compared to that of normal galaxies (gray dots) for objects of increasing intrinsic luminosity. (From *Astrophysical Journal*.)

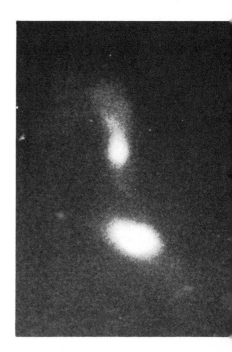

Figure 6.8   Examples of Markarian galaxies: *left,* a pair of Markarian objects (Ma 220 and Ma 221); *center,* a pair consisting of a normal galaxy (top) and a Markarian galaxy (Ma 133); *right,* a clumpy Irregular (Ma 296, bottom). The first two photographs were obtained by J. Heidmann and C. Casini with the 193-cm telescope at the Haute-Provence Observatory. The third plate was obtained by the same two astronomers in collaboration with Lelièvre. (Courtesy of C. Casini.)

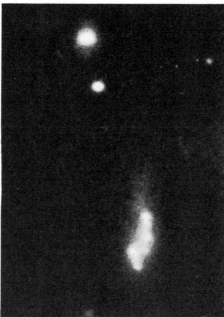

galaxies—the bright nucleus—is quite evident in figure 6.9, which shows the Seyfert NGC 1068 in four plates with increasing exposures. In the first plate we see only the nucleus which, owing to its stellar appearance, can be easily mistaken for a star, especially with short-focus telescopes (of low magnification); in the second plate we begin to discern the underlying galaxy that is quite evident in the third plate, where the nucleus is overexposed; in the fourth plate, obtained with such a long exposure that the whole central region of the galaxy is overexposed, we can make out faint spiral arms that seem to surround the galaxy with a ring of stars.

The high surface brightness of the nucleus allows us to distinguish N and Seyfert galaxies from all others but not from each other. In 1970 Sargent refined the two definitions by taking into account not only the appearance but also the spectroscopic features of the two types of galaxies. At present they are defined as follows:

Seyfert Galaxies

1. Extended, with very bright stellar nuclei.
2. Emission lines of the Balmer series, wider than they appear in the integrated spectra of normal spiral galaxies as a result of the latters' rotation. This condition insures that the only galaxies included are those in which the broadening of the lines is due to explosive phenomena rather than to rotation.
3. Emission line spectrum indicative of a higher excitation level than that normally found in the H II regions of galaxies, namely, exhibiting lines of He II, [Ne V], [O II], [Ne III], [N II], [S II], and [O III].
4. Coexistence of lines of low excitation with lines of high excitation due to these elements.

N Galaxies

1. Extended, with very bright stellar nuclei.
2. Atypical spectrum, with either no emission lines, narrow emission lines, or broadened emission lines in the Balmer series.

From a comparison of the distinguishing features of the two groups, it appears that N galaxies constitute just a broader classification which may include Seyferts as a specific case. As we shall soon see, however, a detailed study of both classes reveals characteristics that in some cases tend to make the distinction between them greater while in others to blur it.

N galaxies have often been confused with another group of extragalactic objects—Zwicky's compact galaxies.[6] The latter, however, constitute a heterogeneous class comprising a great variety of objects. This class, discovered by Zwicky around 1964, consists of galaxies with high-surface brightness and relatively small apparent extension, so that they also often appear like stars. However, not all galaxies in this class seem to have the same physical characteristics (figure 6.10). As Sargent pointed out in 1970, and as later confirmed, most compact galaxies have spectra devoid of emission lines, and the number of stars per unit volume is probably smaller than that found in elliptical galaxies or in the nuclei of spirals, where there is no evidence of nonthermal activity of the type observed in Seyferts. Many of Zwicky's compact systems are nearly normal galaxies that appear compact only because of their great distance. Others are isolated H II regions harboring massive blue stars and so are similar to those in d-ds Markarian galaxies. Finally, a small fraction consists of N and Seyfert galaxies. The classification of compact galaxies is based almost exclusively on their morphological description, which may be the same although due to widely different structures or phenomena. We must be very careful therefore not to confuse this class with the groups we are trying to segregate according to common physical characteristics. This very important task of distinguishing physical features is the only means we have for arriving at a satisfactory interpretation.

Following this approach, we will now proceed to describe the other typical features of N galaxies. Most of them have been found to be powerful radio emitters. In fact, many of these galaxies were first discovered as radio sources, and only later identified as galaxies with optical counterparts.

Another important characteristic of these galaxies is the variability in brightness they sometimes display within very short periods of time. So far brightness variations have been observed in at least twenty-two of them. As in the case of BL Lacertae objects, Seyferts, and quasars, this

6. See *Beyond the Moon*, pp. 261–263.

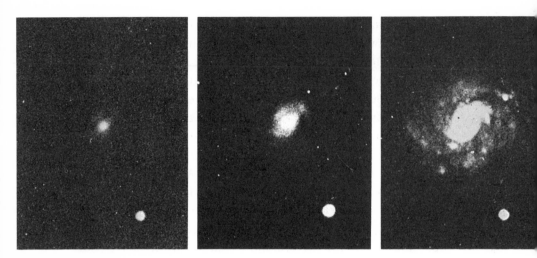

Figure 6.9   Four photographs of the Seyfert galaxy NGC 1068 (M 77) with increasing exposures obtained by F. Bertola with the 1.20-m telescope at Asiago Astrophysical Observatory. The galaxy, which can be mistaken for a star in the first plate, can be clearly seen as an enormous spiral in the last one. (Courtesy of F. Bertola.)

phenomenon indicates a very small component, about one light-year or maybe less in diameter, is the main seat of all the activity.

As we said, N galaxies have something in common with Seyferts, and some of them may in fact be part of this group. An even more marked similarity appears to tie N galaxies in with a class of much more extraordinary objects. In 1973, while performing photometric measurements of a number of N galaxies with Mount Palomar's reflector, A. Sandage found that each of them could be thought of as consisting of two components: a blue pointlike kernel and a red extended component. In the color-color diagram the central source fell among the quasars, while the extended component showed the same color and radial intensity distribution as a giant elliptical galaxy. If the intensity of the blue source increased, the position of the N galaxy in the color-color diagram would shift to the top and left and eventually coincide with that of the quasars. N galaxies appeared to be redder than quasars only because up to that time their emission had been measured for the entire object. In reality, as demonstrated by Sandage, they consist instead of a giant elliptical galaxy with a mini-quasar embedded in its center.

This discovery may be quite significant, particularly if considered together with another finding. As we have known for some time, the lines observed in the spectra of N galaxies exhibit greater red shifts than those observed in Seyferts, and the spectra of quasars show yet greater red shifts than those of N galaxies. If we regard the red shifts as due to a cosmological effect, this means that N galaxies are farther away from us than Seyferts, and quasars even more removed. Consequently, the difference between these three types of objects could be explained as a sort of optical effect: Seyferts, N galaxies, and quasars might simply be objects of the same type that appear unlike because they are observed at different distances. In the first case the galaxy surrounding the nucleus would be quite conspicuous, in the second less so, and in the third it would not be visible at all.

Although this interpretation has been criticized for having some weak points (notably the different spectroscopic features of the objects), it cannot be excluded.

There could also be a second and more intriguing interpretation. In-

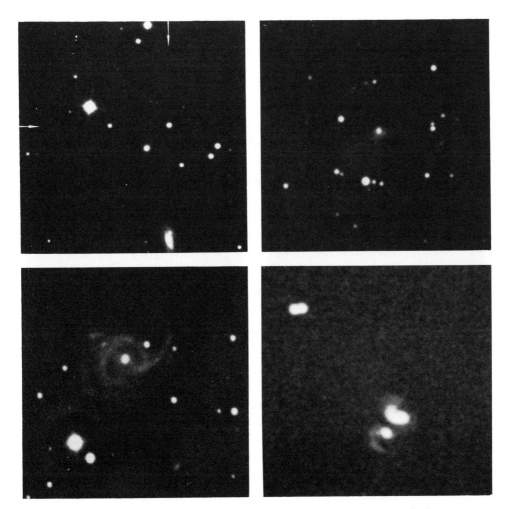

Figure 6.10   These four photographs show four of Zwicky's compact galaxies. This class includes a great variety of objects (from left to right, top to bottom): III Zw 50, a compact stellar galaxy; II Zw 40, a type d Markarian and an intergalactic hydrogen cloud; III Zw 59, an N type or Seyfert galaxy; IV Zw 23, two interacting galaxies. (Courtesy of R. Barbon.)

creasing distances correspond to increasingly more remote epochs. Whereas we observe Seyfert galaxies up to about 30 million years ago, and we see the universe as it was then (practically the same as it is today), we observe N galaxies as far as some hundred million light-years away, and quasars as far as billions of light-years away, that is to say, as they were billions of years ago. It may be that at that time there was a prevalence of quasars, which subsequently evolved into N galaxies; today we would be seeing those few N galaxies that survived until some hundred million years ago. Finally, as we come to today's universe, we find the latest product of evolution—Seyfert galaxies—which might represent either a transient and short-lived phase in the evolution toward normal galaxies or an exceptional phenomenon of short duration, perhaps recurring, that only takes place in some peculiar galaxies.

This interpretation has a serious drawback. It requires that galaxies should have evolved rapidly in the recent past (the last billion years). But on the basis of various findings, it is our current belief that the evolution of the universe was faster in the initial stages, in the first billion or millions of years.

SEYFERT GALAXIES

Let us now examine Seyfert galaxies in greater detail. We shall begin from s-sd Markarian galaxies and recall that this class includes all galaxies with ultraviolet excess and starlike nuclei. This class, which very nearly corresponds to that of N galaxies, can be divided into two groups. One group consists of galaxies with bright nuclei whose spectra show narrow emission lines. In the color-color diagram (figure 6.1) they fall in the proximity of the line corresponding to the black body, and their radiation is entirely produced by thermal processes. The second group comprises galaxies that also have nuclei of stellar appearance but emit nonthermal radiation, whose spectra exhibit broadened emission lines. The former are called "bright nucleus" galaxies, or BN galaxies; the latter are Seyfert galaxies (their definition was given earlier when we compared them with N galaxies).

Seyfert galaxies can in turn be divided into two groups: type I Seyferts, whose spectra exhibit permitted lines broader than the forbidden lines; and type II Seyferts, in which both permitted and forbidden lines appear of the same width. This difference is often determined by comparing the width W of the H_β hydrogen line (permitted) with that of the nearby line of [O III] (forbidden). If a Seyfert galaxy has a spectrum where $W_{H_\beta} > W_{[O\ III]}$, it belongs to type I. If the spectrum shows $W_{H_\beta} \approx W_{[O\ III]}$, the galaxy is a type II Seyfert. It goes without saying that we are speaking of emission lines.

The two types of galaxies fall in different regions of the color-color diagram (figure 6.11). All radio-emitting Markarian galaxies that have turned out to be Seyferts belong to type II. It is thought that these facts may be indicative of a significant difference between the two types, and this belief has been confirmed by a recent discovery that we will discuss shortly.

The causes of this difference almost certainly reside in the nuclei, where the observed emission lines are formed and explosive activity takes place. Moreover, if we disregard the nuclei, almost all Seyferts would look like normal spiral galaxies and could be confused with all the others. Consequently, it is mainly the nuclei we must examine, and we shall now attempt to take a closer look at them, or, to be precise, at their central regions.

To begin with, it has been found that the nucleus is very small as compared to the size of the whole galaxy. According to measurements performed outside the earth's atmosphere by M. Schwarzschild, the nucleus of NGC 4151 appears to be no larger than 7 parsec. Seven parsec, or 23 light-years, are really nothing if we consider that the diameter of a galaxy is at least a thousand times greater. But the nucleus of NGC 4151 is actually even smaller. On the basis of its variability it has been computed that it cannot be larger than 0.05 parsec, or only 2 light-months. The mechanism responsible for all the peculiarities observed in Seyfert galaxies must be contained therefore within a very small volume.

This strange machine does not seem to be that simple. Its spectral lines show that it must contain gas, and the distribution of energy in the continuum reveals the presence of a nonthermal source. In addition, the nuclei of several Seyferts must contain large amounts of dust. This is confirmed by observation of significant infrared radiation, presumably due to clouds that absorb the enormous amount of energy radiating from the inner regions and re-emit it outward at longer wavelengths.

This is the most immediate interpretation but not the only one. In the summer of 1975 T. F. Adams and D. W. Weedman studied the correlation between the intensity of some lines they had observed in the spectra of the nuclei of 32 Seyferts and the characteristics of the continuum obtained from UBV photometry. They concluded that only the nuclei of Seyferts II contain significant amounts of dust. According to these authors, the peculiar properties exhibited by the nuclei of Seyferts II can be explained by the presence of very large numbers of extremely hot, bright, and reddened stars.

The nuclei of Seyferts I seem to be substantially different. They are not reddened by dust in any appreciable amount, and in this case the infrared emission may be due to an extension of the continuum of nonthermal origin. This recent discovery, confirmed in the spring of 1976 by observations carried out by W. A. Stein and D. W. Weedman, points to a basic difference between Seyferts I and II and shifts the problem to be solved to the galaxies of the former type. Therefore we shall examine them a little further. As we have learned earlier from their definition, these are the galaxies where permitted lines appear broader than forbidden ones.

Observing their emission lines over the span of a few years, it appears that they do not vary significantly with time, with the exception of three galaxies: NGC 1275, NGC 1566, and IC 450. Since the continuum varies erratically, we conclude that the region emitting the lines is far larger than the region emitting the continuum. Furthermore, L. Woltjer had already suggested in 1959 that the substantial width of the emission lines might be due to a great dispersion of the velocities in the radiating gas. This interpretation was confirmed by the observations carried out by M. F. Walker in 1968 on the nucleus of NGC 1068 and by M. H. Ulrich in 1973 on the nucleus of NGC 4151. In both cases the nuclei were found to be surrounded by a number of discrete clouds with diameters of a few hundred parsec and masses ranging from 1 million to 10 million $M_\odot$, which were moving away from the center with velocities as high as 600

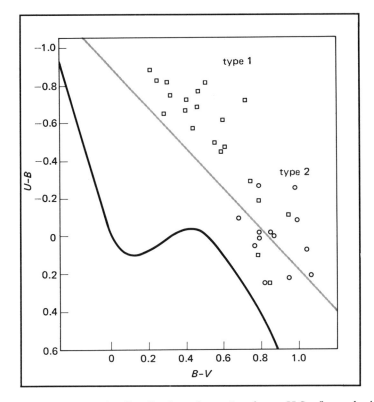

Figure 6.11   The distribution of type I and type II Seyfert galaxies in the color-color diagram. The two groups appear well separated, although some galaxies of type I fall in the group of type II.

km/s. The age of these clouds, computed from their expansion, turned out to be between 100,000 and 1 million years.

Both the forbidden lines (narrow) and the central region of permitted lines (broad) appear to originate from these clouds; the wings (or outer parts) of the permitted lines, like those observed in Seyferts I, would originate from a much smaller and much more turbulent region. A good example is the Seyfert galaxy IC 4329 A whose spectrum, according to M. J. Disney, exhibits the forbidden lines of [O III] with a width corresponding to only 1,500 km/s, while the wings of the H_α line (permitted) correspond to an expansion velocity of 13,000 km/s. The different behavior of permitted and forbidden lines can be explained if we assume that the light corresponding to the wings of the permitted lines originates in a small volume where the gas attains such a high density that it precludes the formation of forbidden lines.

All these observational results give us a pretty good idea of the nucleus of a Seyfert of type I. According to a model which on the whole is quite satisfactory (figure 6.12), at the exact center of the galaxy is a small but very intense source that emits nonthermal radiation. This source is surrounded by a region about 0.2 parsec in diameter (two and one-half light-months) which contains dense clouds expanding at speeds of about 3,000 km/s and emitting radiation in the permitted lines, such as the Balmer hydrogen series. At a distance of about 400 parsec (1,300 light-years) there are tenuous clouds that expand more slowly, at speeds of about 300 km/s, and emit the forbidden lines.

A number of hypotheses have been proposed to account for the excitation of the emission lines: intense ultraviolet radiation emitted by the nonthermal central source, shock waves, or supersonic wind—all of which would be associated with an explosion.

The results we have presented so far are of fundamental importance. However, Seyfert galaxies exhibit also some other interesting features that we cannot overlook, even though it is not yet clear how they fit in with the picture we have drawn. Instruments installed on board the Uhuru satellite, which went into orbit on December 12, 1970, disclosed that at least two Seyferts—NGC 1275 and NGC 4151—emit X rays. In

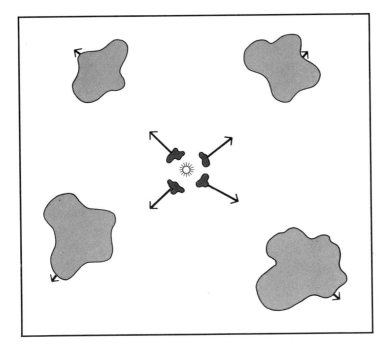

Figure 6.12   A schematic representation of the nucleus of a Seyfert galaxy. The model consists of three components: a small central source of nonthermal radiation; a larger region of about 0.1 parsec in radius in which dense high-velocity (3,000 km/s) clouds emit permitted lines; and finally an outer region of about 400 kpc diameter in which lower-velocity (300 km/s) and lower-density clouds emit forbidden lines. (From S. van der Bergh.)

addition, Seyfert galaxies are also radio emitters. In the nuclei of the nearest and brightest Seyferts observed to date, a small radio source has been found with an intensity from 10 to 1,000 times higher than that observed in normal spiral galaxies. No Seyfert galaxy, however, exhibits a double radio structure such as that found in most radio galaxies and quasars. This is in fact the one great difference between Seyferts and quasars.

The diameters of the radio sources, like those of the nuclei, appear relatively small. This conclusion has been supported by measurements made in recent years by means of radiointerferometry, by coupling very distant radio telescopes so as to obtain a single instrument with an intercontinental baseline. This technique produced additional results. It was found that NGC 1275 harbors a central radio source only 0.02 arc seconds in diameter, which in turn has been resolved into a number of smaller compact sources of variable intensities, each with a diameter smaller than 0.001 arc seconds, or one light-year. These pointlike sources are embedded in a sort of halo whose energy content is about one million times greater than that of the smaller sources. The halo is believed to have formed from the combined effects of several explosions of the type we are currently observing in the galaxy.

One cannot help wondering what strange machine is hidden at the center of that galaxy and others similar to it. Such prodigious emission of energy and matter from a region that appears to be shrinking, the more we study it, poses questions to which we have no answers.

Despite this mystery, it is also true that Seyfert galaxies are no more strange or mysterious than some of the objects we have discussed earlier, such as Lacertids or N galaxies. Moreover, the most difficult problem is to explain the connection between all these objects and whether or not all their differences and similarities have any evolutionary meaning. A discussion of the possibility is perhaps premature. In 1975, however, S. van der Bergh attempted an empirical classification which suggests interesting correlations between extragalactic objects with excess UV. We reproduce it here for illustrative purpose (figure 6.13). According to this scheme, Seyfert galaxies appear to assume a much greater importance in

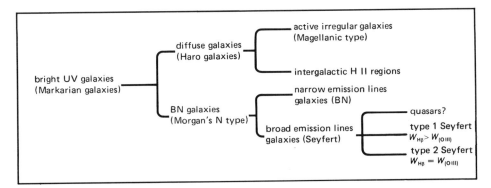

Figure 6.13   A classification scheme for extragalactic objects with UV excess proposed by van der Bergh. (From *Journal of the Royal Astronomical Society of Canada.*)

that they are connected to the most mysterious objects of all: the quasars. And it is the quasars that we will examine in the next part of our journey.

QUASARS

At the time they were discovered, quasars appeared to be mysterious and unique objects. Although the mystery persists, quasars do not seem to be quite so unique now. Rather they appear to be related to some of the puzzling extragalactic objects we have just encountered. We shall now present some of the latest results in this field, particularly from the standpoint of the relation between quasars and other active galaxies.

The main characteristics of the quasars and the various interpretations that have been proposed are very well known by now.[7] They are objects of stellar appearance, which may be radio emitting or radio quiet; of variable brightness and radio emission; with very peculiar spectra exhibiting an excess in ultraviolet and infrared emission, few absorption lines, and highly red-shifted emission lines. If we consider the red shift as due to the general expansion of the universe—which was the interpretation given for the red shifts observed in the spectra of normal galaxies—we find that quasars must be located at enormous distances: from 1 to 10 or 15 billion light-years. Taking into account their apparent magnitudes, it turns out that their intrinsic luminosities must be far greater than those of the brightest galaxies.

As a rule their absolute visual magnitude is about -24, and their emitted energy is about 10^{46} erg/s. But there are some exceptional cases where even these values have been greatly exceeded. We shall recall two events brought to our attention in 1975 in which the emission attained extraordinary values even for quasars.

The first one is PKS 1510-089, with $z = 0.361$. Old Harvard plates dating from 1899 revealed that its brightness at minimum was magnitude 17.8, corresponding to an absolute magnitude $M_b = -23.9$, which is consistent with the typical luminosity of quasars. In 1948, however, it

7. See *Beyond the Moon*, pp. 301–310.

flared up to magnitude 11.8. No quasar had ever shown such a large variation. It was computed that during this outburst, lasting for about twelve days, it emitted a total energy equal to or greater than 10^{53} erg. During that period it was a millionfold brighter than a type I supernova. Furthermore, the rapidity of the variation showed that this fantastic amount of energy had been emitted by an object whose diameter could not exceed 1,700 AU, or a mere seventeen times larger than the solar system.

Even more remarkable is the quasar 3C 279, with $z = 0.536$, which attained $m_{pg} = 11.3$, corresponding to an absolute photographic magnitude $M_{pg} = -31.4$. With the exception of the farthest quasars (for which it is better to suspend judgment since the computation of distances and therefore of absolute magnitudes is critically dependent on the deceleration parameter, q_0, which is still not well determined[8]), this quasar is currently the most luminous object known in the whole universe.

On the basis of everyday experience, we cannot even begin to understand such brilliance, but we can perhaps gain a faint idea of it by a simple comparison. Suppose our planet, together with the solar system, were to approach that quasar at a standard distance of 10 parsec (32.6 light-years). At that distance, and in the period of its maximum brightness, it would appear to us a hundred times brighter than the sun and would totally dominate our night sky. Of course, at this point we could call it a night sky only because of the absence of the sun. Our eyes would have to become accustomed not to sunlight but to the blinding light of the quasar, radiating from a point rather than from a disk. An unbelievable radiance would wash over our landscapes and blot out the light of the stars much more effectively than the sun is able to do during the day.

At a certain time, because of the earth's rotation, the quasar would set below the horizon, and the sun would rise ushering in the new day. But as compared to a "night" illuminated by the quasar, daylight would seem almost as pale as moonlight. Naturally during the course of the year there would be times when both the sun and the quasar are above the

8. See the appendix.—Trans.

horizon. While the quasar would continue flooding earth and sky with its light, the sun would appear as a golden disk that our eyes, accustomed by then to the light of the quasar, could behold as easily as today's setting sun. Then, after the sun and the quasar have set, there would come a true "night," and we would finally see the stars. The quasar would rise again, sometimes more luminous, sometimes less so, owing to its variability, but always far brighter than the sun except for the rare instances when, having dimmed to the minimum at about absolute magnitude -26, it would appear to be about the same brightness as the sun.

This is what our sky would look like if we had a quasar like 3C 279 not in place of the sun but 2 million times farther away.

The scenario we have just described is based on the assumption that the red shifts observed in quasars are of cosmological origin. This is the opinion of most, but not all, scientists. However, some findings of the last few years have cast doubts on this theory and reopened a controversy that had begun at the time the quasars were first discovered.

In 1973 three separate instances were reported of a quasar that appeared to be in the vicinity of a small group of galaxies. The red shifts of the galaxies in all three groups could not be larger than 0.6, because for a higher value the galaxies would be too far away to be detected with our telescopes. On the other hand, the values of z in the three quasars were computed at 1.27, 2.17, and 0.73, respectively. They are all too high if we assume that the quasars are associated with the galaxies and are therefore at the same distance.

There is a second, more convincing case of two pairs of quasars where the two components appear so close together as to exclude the chance of a purely random superposition. The components of the first pair are 35 arc seconds apart and exhibit red shifts of 0.55 and 1.17, respectively. In the second pair, the quasars are only 4.5 arc seconds apart and have red shifts of 0.436 and 1.901. In both cases the cosmological interpretation of the red shifts would result in an enormous distance between the two objects, in contrast to their small angular separation.

These puzzling differences in the red shifts of objects that appear close, or even connected, to each other are by no means rare, and they constitute one of the most fascinating and controversial subjects of current extragalactic astronomy. We shall talk about it again. Right now, for the sake of impartiality, we should point out that the cosmological interpretation of the red shifts favored by the majority of astronomers has also found confirmation in recent observations. In 1973 R. L. Brown and M. S. Roberts found that the radio spectrum of the quasar 3C 286 showed the 21-cm hydrogen line in absorption, with $z = 0.69$. The optical spectrum of the quasar revealed a greater red shift, $z = 0.85$. Hence the quasar must be much farther away than the region containing the hydrogen, which causes the absorption in the 21-cm line and by mere chance must be located just behind it. Evidently this region corresponds to a galaxy too far away to be seen. If we assume that the galaxy's red shift is of cosmological origin, it must also be so for the quasar which is surely much farther away.

A second proof, of a statistical nature, is based on the presence of quasars in clusters of galaxies. Several quasars have been observed in clusters of galaxies. In view of the distribution of quasars in the sky, it has been calculated that their random association with clusters should be less frequent than that actually observed. This prediction seems to be confirmed by the fact that the red shifts of the quasars observed in clusters are consistent with the average red shifts of the nearby galaxies.

A. Sandage produced a third proof by showing that N galaxies could be regarded as consisting of a giant elliptical galaxy and a mini-quasar. By constructing Hubble's diagram for the elliptical galaxy component of twelve N galaxies, Sandage showed that the same slope, dispersion, and zero point could be obtained as in the diagram drawn for radio galaxies. He thus proved that the red shifts of the N galaxies were purely cosmological and that the red shifts of the mini-quasars embedded in them must be cosmological as well.

This result brings us back to a possible connection between quasars and certain types of galaxies. We have just seen a possible connection with N galaxies and, a little earlier, with Seyferts. In addition, another interesting correlation was suggested in 1971 by D. Lynden-Bell. According to this author, the observed quasars are the exponential tail (where luminosities are highest) of a population consisting of a very large number of fainter objects that generally escape detection. In his opinion, not all of these fainter objects have escaped detection; they would be partly N galaxies and partly Zwicky's compact galaxies (or at least some of them). The first group would constitute the extension to faint magnitudes of the radio-emitting quasars and the latter of the radio quiet.

This hypothesis appears very convincing. Furthermore, the existence of smaller or fainter quasars that we see as the bright nuclei of galaxies of low surface brightness also suggests that the brighter but more distant quasars might be at the center of galaxies we cannot see. In other words, it is possible that *all* quasars are the nuclei of peculiar galaxies (N, Seyfert, and compact) that we can only see in their entirety when they are closer to us, that is, in the case of the less luminous systems which are also more frequent. At great distances therefore we would see only the brightest objects of the entire population of peculiar quasar-galaxies, and of these objects what we see is only the starlike nucleus.

In 1972 J. Kristian undertook a search for galaxies associated with quasars. Unfortunately, this is not as simple as it may appear at first sight. The use of a large telescope and long photographic exposures do not always guarantee that we will see more. In order to show how difficult these investigations are, we will describe not only his results but also the method he used. It must be kept in mind that research in this field is often carried out at the limit of our instrumental capabilities and requires a great deal of acumen on the part of the researcher. Often, after much labor, it is still difficult to reach definitive, yes-or-no conclusions, which would be so useful in developing a theory but are very hard to attain in experimental work.

Our goal, at least in principle, is very simple: discover the galaxy that surrounds the quasar located at its center. In practice, it is not so simple. The underlying galaxy can be detected on a photographic plate only if the diameter of its image is greater than that of the quasar's image. The diameter of the latter is determined by its brightness, while the diameter of the image of the galaxy depends on the galaxy's intrinsic diameter and distance. Let us assume that all the galaxies we are looking for have the same diameter, which is not an unlikely possibility. In order to determine for which quasars it might be possible to detect the underlying galaxies with a given telescope (for instance, the 5-m telescope at Mount Palomar), it is necessary to calibrate the size of the quasar's image as a function of its apparent magnitude. In addition, the apparent diameter of the galaxy, photographed with the same telescope, must be determined as a function of its distance measured by its red shift.

As we said, the underlying galaxy can be detected only if the image of the central quasar has a smaller diameter than that of the galaxy itself, that is, if the quasar is rather faint and the entire object is sufficiently close. Following this reasoning, and having performed the two necessary calibrations, Kristian observed that Hubble's diagram (where apparent magnitudes are plotted on the abscissa and the red shifts on the ordinate) could be divided into three regions: one region corresponds to diameters of the galaxy greater than twice that of the central quasar's image; another region is that in which the diameter of the galaxy is between one and two times that of the quasar; and, finally, there is a region in which all the galaxies appear smaller than the image of the quasar and therefore cannot be detected.

Kristian selected twenty-six quasars suited to his purpose and photographed most of them with the 5-m telescope of Mount Palomar. In all cases where detection was technically possible and good plates could be obtained, the galaxies were indeed discovered. The diameters of these new galaxies associated with quasars turned out to be of the same order of magnitude as those of N galaxies or of the brightest galaxies in clusters.

We can now confidently say that not only are there quasarlike objects at the center of certain galaxies but that, whenever possible, we have found galaxies associated with quasars. The last conclusion brings us back once again to the possibility that all these objects may represent an evolutionary chain that begins with quasars and ends with N galaxies. Following this line of thought, one could think that normal quasars are much brighter than the equivalent objects (mini-quasars) observed at the center of N galaxies and therefore that the galaxy believed to surround a quasar may be comparatively much fainter and much harder to photograph than an N galaxy.

Kristian's results, published in 1973, have not been pursued any further, but it must be pointed out once again that this type of research is very difficult and is feasible only for a limited number of quasars. Even in these few cases the detection of the underlying galaxy really strains the capability of the world's largest telescopes.

A recent discovery has revealed a correlation between quasars and another group of extragalactic objects—the Lacertids—that is to say, the very objects we encountered at the beginning of this chapter. In studying two BL Lac objects, as well as nine possible candidates, J. T. Pollok found a correlation between the amplitude of the optical variations and the radio spectral index at 5 GHz. This result was published in June 1975 together with the even more significant finding obtained independently by P. D. Usher. In the latter case the sample consisted of a greater number of objects (37) and also included quasars and N galaxies. The result of Usher's work is an analogous relation in which quasars differ from the other objects only because they occupy the lower left-hand corner of the resulting diagram (figure 6.14). Thus the Lacertids also appear related to quasars.

At this point we will call a halt to our journey among the puzzling objects that we must somewhat hesitatingly call galaxies. If we consider the prodigious amount of energy they continuously radiate into space, their compact size, variability, and violent explosive activity, there is no doubt that all these objects—Lacertids, quasars, Markarian, N and Seyfert galaxies—appear truly extraordinary. Only a few astronomers still believe that some of these objects may not be extragalactic. But if they do belong to the world of galaxies, as generally accepted, their relation to the beautiful normal elliptical or spiral galaxies is certainly not clear at all. Perhaps they coexist with galaxies, without having any relation to them. But the most likely hypothesis, and the one that is continuously reinforced by observations, is that they represent a certain facet of the world of galaxies corresponding to very ancient or very rapid phases of their evolution. Thus these monsters appear to be the key to the interpretation of the evolution of galaxies and of the universe itself.

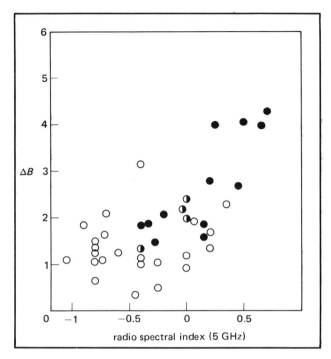

Figure 6.14 The amplitude of the optical variations (in B magnitudes) as a function of the radio spectral index for quasars (open circles), weak emission lines objects (half-filled circles), and sources with continuum emission (filled circles). BL Lac objects fall among the latter. (From *The Astrophysical Journal.*)

ANOMALOUS RED SHIFTS

FACTS

While discussing the quasars, we came across two "odd couples" in which the spectrum of each component showed an appreciably different red shift. Assuming that the red shift was due to the cosmological effect of the expansion of the universe, the observed difference told us that the quasars in each pair were at different distances and appeared close to each other only for reasons of perspective.

In those two cases a random superposition was possible, although rather unlikely. Things would have been quite different, however, if in addition to their apparent proximity there had been other indications that the objects were actually near each other and therefore at the same distance from us. For instance, if a physical tie between the two quasars could be proved with certainty. It now appears that such cases might actually exist.

One of the most interesting and controversial is the case of NGC 4319—a rather peculiar spiral galaxy located near the Markarian quasar 205 (figure 6.15). The two objects appear to be moving away from us at radial velocities of 1,800 and 21,000 km/s, respectively. Thus the first should be at a distance of 33 Mpc (108,000,000 light-years) and the second at 382 Mpc (1,250,000,000 light-years). Hence the distance between the galaxy and the quasar should be 349 Mpc, or more than 1 billion light-years, which is far greater than the distance between us and the galaxy. We have reproduced a picture of the objects obtained by H. Arp with the 4-m telescope at Kitt Peak National Observatory by superimposing three plates. The two objects appear connected and therefore, according to Arp, must be considered at the same distance from us. Arp concludes that in one of the two objects the observed red shift *cannot* be cosmological.

There has been considerable discussion over this case. Among other things, not all those who have seen the picture are convinced that there is an actual connection. Some have pointed out that the dark region connecting the quasar to the outer part of the galaxy may be a photographic effect due to the superposition of the dark areas corresponding to the

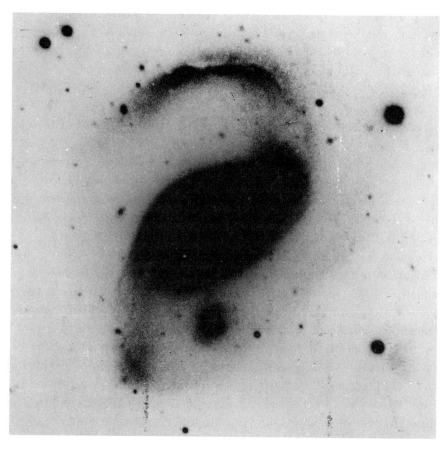

Figure 6.15   The pair (real or due to coincidence) consisting of the galaxy NGC 4319 and the Markarian object Ma 205 (the round object below the galaxy) in a photograph obtained by H. Arp. (Courtesy of H. Arp.)

peripheral regions of the objects. In conclusion, it seems to be a matter of opinion, and the reader can judge for himself by examining figure 6.15, which reproduces the original negative obtained by Arp.

There are additional examples. A remarkable photograph obtained by Arp with the same technique shows that the Seyfert galaxy NGC 7603 has a companion which appears connected to the main object by a curved filament, even though their radial velocities are 8,800 and 16,900 km/s, respectively. A third instance is that of NGC 772, with $v = 2,430$ km/s. Close to it are three smaller objects that, according to Arp, are satellite galaxies connected to the main one. One of them has the same radial velocity as the main object ($v = 2,450$ km/s), but the radial velocities of the other two have been computed at 20,200 and 19,700 km/s. In addition, there are other instances of small groups of galaxies in which one or two of them have radial velocities substantially different from all the others. We will mention here only VV 159, VV 43 (Zwicky's triplet), VV 115 (Seyfert's sextet, figure 6.16), and VV 172 (figure 6.17).

The last chain of galaxies, in particular, consists of five objects, four of which have radial velocities of about 16,000 km/s, while for the fifth one $v = 36,880$. It does not seem possible that this galaxy (second from the top in figure 6.17) should actually be very far away from the other four and should be part of the chain by mere chance, falling precisely in an empty space. On the other hand, if we assume that all five galaxies are physically connected and at the same distance from us, we must also assume that the red shift of the second one is not cosmological. But then, what is the red shift due to in the various cases we have just discussed? And if it could be proved that some red shifts are not cosmological, how can we trust all those we have measured in the last half a century? It is on the basis of the red shifts that we have computed the distance of the farthest extragalactic objects and discovered the mind-boggling concept of the expansion of the universe which is the cornerstone of the Big-Bang theory of modern cosmology. The anomalous red shifts, which we would not know how to explain, seem to undermine all our carefully constructed theories and force us into a general revision of current cosmology, at the same time depriving us of the basic technique for probing the universe. These problems are fairly dramatic, and as long as they

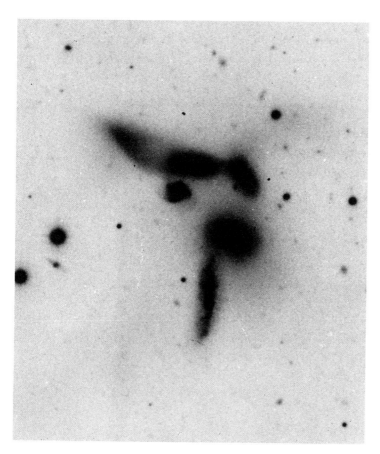

Figure 6.16   Seyfert's sextet in a photograph obtained by H. Arp with the 122-cm Schmidt telescope at Mount Palomar in a 90-minute exposure. The radial velocities of four of the galaxies are 4,500 km/s; the fifth is 20,000 km/s. (Courtesy of H. Arp.)

threaten our observational results like the proverbial sword of Damocles, the future of cosmology seems somewhat uncertain.

There are several problems that need to be solved in order to explain this enigma. In the first place we must find out if the anomalous red shifts are so numerous as to invalidate Hubble's law. On the other hand, even if the extreme cases (like those we have just discussed) should be few, there would still remain the doubt that perhaps in a very large number of other cases the effect was too small to be noticeable but was nevertheless present and capable of confusing our results. Specifically, the red shift we observe in a galaxy or quasar could be due to the sum of three components: one cosmological, one due to the radial component of the galaxy's motion, and the last one caused by an unknown factor inherent in the galaxy itself. The second component has always been considered negligible, at least for large distances. In order to rely with confidence on the cosmological component, we would also have to disregard the third one, that is, the anomalous red shift. But first of all we must go back to the fundamental question: Does the anomalous red shift really exist? From what we said earlier, the reader will appreciate that this problem is very tricky and extremely hard to solve on the basis of our current knowledge. The only way is to tackle each case separately; and we will examine only one here—the one most widely studied and discussed.

STEPHAN'S QUINTET

On September 23, 1876, E. Stephan, an astronomer at the Marseilles Observatory, discovered a tight group of four small nebulosities that were barely visible with the 80-cm refractor of that Institute. Eleven years later the four objects were listed in Dreyer's well-known catalogue as NGC 7317, NGC 7318, NGC 7319, and NGC 7320. It was subsequently discovered that NGC 7318 really consisted of two objects, and the quartet became a quintet. With the advent of large telescopes, the group was photographed at Mount Wilson Observatory. Astronomers there found that the five small nebulosities were in fact five large galaxies which, judging from the red shifts observed in their spectra, appeared to be very far away—360 million light-years, according to the latest mea-

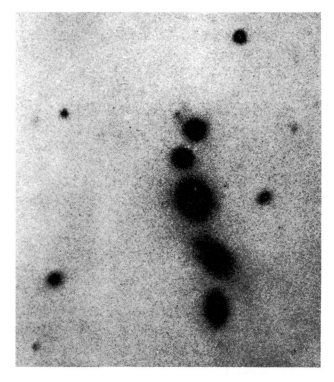

Figure 6.17   The chain of galaxies VV 172 consisting of five objects, four of which have very similar red shifts. (The Hale Observatories.)

surements (figure 6.18). The distance was computed from the radial velocities which were 6,700 km/s for three of the galaxies and 5,700 for the fourth; the radial velocity of the fifth—NGC 7320—had not been determined. It was computed in 1960 by the Burbidges who, to their great amazement, found it to be only 800 km/s. Hence NGC 7320 could be no farther away than 40 million light-years and appeared close to the other four only for reasons of perspective. The chance of this random superposition was estimated at one in a thousand. Rather than accept such an unlikely possibility, many astronomers choose to believe that the region of Stephan's quintet had been the seat of a tremendous explosion that had propelled a whole galaxy—NGC 7320—in our direction.

H. Arp, who had already begun to consider the possibility that there might be anomalous red shifts, thought that Stephan's quintet could be just one of these cases. However, the anomalous red shifts he had found until then were all in excess, much higher than the values expected from the cosmological effect. Consequently, the normal red shift corresponding to the actual distance of the quintet had to be that of NGC 7320, while the other four had to be anomalous.

Further examination of the region (figure 6.19) disclosed that at only half a degree from Stephan's quintet there is a large galaxy, NGC 7331, with a radial velocity of about 800 km/s, the same as that of NGC 7320. Moreover, the large galaxy has six smaller companions located in the exact opposite direction from the quintet, with radial velocities of about 6,000 km/s, about the same as the other four members of the quintet.

Arp then advanced a very interesting hypothesis. He proposed that both Stephan's quintet and the group of satellite galaxies were expelled in diametrically opposite directions by the large galaxy NGC 7331. As a result of the explosion, all these galaxies now have anomalous red shifts, with the exception of NGC 7331, the parent galaxy, and NGC 7320 which has returned to a purely cosmological red shift, possibly because it had been expelled earlier.

Subsequently, Arp employed the world's best instruments and all of his ability as an observer to prove his hypothesis and the existence of anomalous red shifts. The results of his efforts were extremely interesting. He found an excess of radio sources in the vicinity of NGC 7331 and

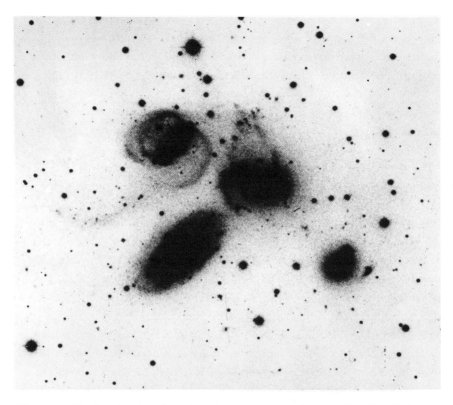

Figure 6.18   Stephan's quintet in a deep exposure plate obtained by H. Arp.
The galaxy appearing at the center really consists of two very close galaxies. (The
Hale Observatories.)

on the side of the quintet. Combining six photographs of the region obtained with the Schmidt telescope of Mount Palomar, he discovered a nebula that appeared to connect NGC 7331 with Stephan's quintet. He photographed with a sophisticated technique a number of H II regions in NGC 7320 and in NGC 7318 B and found that the two galaxies had to be at the same distance. Finally, he discovered a tenuous filament that extended out from NGC 7320 toward a nebulous region found to have a radial velocity of 6,000 km/s.

Despite these intriguing results, from 1970 on evidence against Arp's hypothesis began to mount up. G. A. Tammann pointed out that the beautiful photographs obtained by Arp with Mount Palomar's 5-m telescope showed that NGC 7320 was partially resolved into stars, while the other galaxies were not. Hence the latter had to be much farther away. From spectroscopic observations of the entire region, C. R. Lynds found that three galaxies (NGC 7320, NGC 7331, and NGC 7343) had radial velocities clustering about 1,000 km/s, while the remaining four galaxies of the quintet, as well as seventeen other uncatalogued galaxies in the region, had radial velocities of about 7,000 km/s. Thus it seemed rather untenable that over twenty galaxies should have anomalous red shifts in order to belong to a group defined by only three objects. It was more reasonable to think that there existed a distant, more numerous group over which three closer galaxies appeared superposed and that one of the three, NGC 7320, happened to be located just in front of a group of four, thereby forming Stephan's quintet.

This was the situation in 1971 when there occurred an event that brought forth unexpected help and totally unforeseen evidence: a supernova ignited in one of the four controversial galaxies. The star appeared in NGC 7319 and was discovered by L. Rosino at the Asiago Observatory. At the time of discovery, the supernova was already past maximum, but, as we explained in the chapter on supernovae, Rosino and his collaborators had previously determined that the light curve of all type I supernovae is exactly the same. Rosino drew the light curve of the new star starting from the time of discovery and, by fitting it to the standard curve, computed the date of maximum and the apparent magnitude at that time. Since the value of the absolute magnitude at

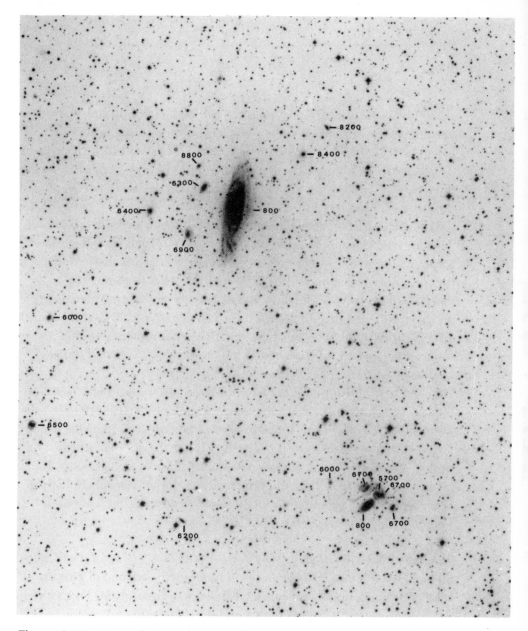

Figure 6.19 A large-field photograph of Stephan's quintet and other galaxies in the region. The numbers shown near each galaxy are their red shifts expressed in km/s. NGC 7731, the large galaxy at the top of the figure, and NGC 7320 in Stephan's quintet, at the lower right, are the only galaxies in the figure with a red shift considerably different from all the others. (Palomar Sky Survey.)

maximum is the same for all type I supernovae (-18.6), Rosino was able to compute the distance of the supernova and therefore that of the galaxy NGC 7319 in which it had appeared. This distance was 138 Mpc, in any case, no less than 102 and no more than 175 Mpc. This value was quite consistent with that of 111 Mpc obtained in the hypothesis of a purely cosmological red shift.

This result should have settled the matter. Instead, astronomers are still debating over the anomalous red shifts of the four galaxies and whether Stephan's quintet is really a quintet or a quartet. Arp has objected that the supernova might have been abnormal, but it has been pointed out that the abnormality would have had to be truly exceptional to explain the difference of 5 magnitudes necessary to place NGC 7319 at the same distance as NGC 7320. Against the negative proof provided by the supernova, the only valid argument left on Arp's side is that of the H II regions.

In late 1972 a group of French radio astronomers used a new technique to measure the distance of galaxies and found NGC 7319 to be at a distance of 22 Mpc, or a little less than twice that of NGC 7320, for which they confirmed a distance of 12 Mpc. As a result the question still appears to be controversial. However, as one of the astronomers who obtained the new result, P. Chamaraux, said, "majority rule does not apply here, because the argument of the supernova carries by itself as much weight as the other two combined."

A DEBATE AND A TRIAL

Stephan's quintet is a good example of how much work and discussion have gone into a single case which has still not been resolved. To give a definite answer to the fundamental question of whether anomalous red shifts really exist, a similar effort is required for each of the other cases, which are generally even harder to study. In this effort, as Arp himself has remarked, the advocates of anomalous red shifts are in a better position than their opponents. While the former need only find a single uncontestable case, the latter must criticize and refute the anomalous cases one by one on the basis of established effects or phenomena. This "detective story" atmosphere not only spurs researchers to intensify their efforts but also provokes continuous discussion aimed at establishing the reality of the phenomenon and understanding its import, should it be positively confirmed.

On December 30, 1972, the two principal exponents of the opposing viewpoints, H. Arp and J. N. Bahcall, confronted each other in a friendly debate during which each presented his point of view and listened to the other's objections. The minutes of this meeting were later published in a volume together with reprints of the most important articles and scientific papers that had previously appeared in various magazines and journals in support of either theory.[9] Although no conclusion was reached, there is no better source for an objective examination of the problem.

Less well known, but much more lively and comprehensive, was a series of debates carried out at the Meudon Observatory from January to May 1973, in which a large number of theoretical and observational astronomers participated. These discussions were made public in an internal publication of Meudon Observatory as minutes of the "Tribunal of the Anomalous Red Shifts." Once again, the general consensus that emerged from these meetings is that the existence of anomalous red shifts has not yet been proved, even though some cases are undeniably puzzling. At any rate it does not appear likely that the anomalous cases will change the general principle of the expansion of the universe. These cases are indeed very few as compared to the extremely large number of galaxies that satisfy it with great precision.

The more important question at the moment seems to be a different one. We have repeatedly stated that there is no explanation as yet for the anomalous red shifts. This does not mean that conjectures have been lacking. One of the theories involves a possible photon-photon interaction; another even bolder theory postulates that physical constants, such as the mass of the electron, might vary from one end of the universe to the other. All these solutions would present a terrible threat to the very foundations of current physical theories. The hypothesis proposed to

9. G. B. Field, H. Arp, and J. H. Bahcall, *The Redshift Controversy* (Reading, Mass.: W. A. Benjamin Inc., 1973).

explain the exceptionally large red shifts of quasars by a noncosmological effect, and adopted by Arp himself as a working assumption, appears to us as the most dazzling and far reaching of all. We are talking of the almost fantastic possibility that we are dealing with "white holes," from which matter and energy coming from another universe, or transferred from different places and times of our own, would spew out into our universe. The red-shift excess would thus be due to the extreme youth of the newly appeared matter and would be explained in a noncosmological way both for certain objects associated with galaxies, from whose nuclei they would have just been expelled, and for quasars, which would no longer be as far away as they are now believed to be according to the classical interpretation.

Thus we have come back again to the black holes, or the fragments of the original singularity that exploded at the beginning of the universe. They beckon us into a world of new physical laws that we could explore only by abandoning our laboratories and studying matter in the most remote places beyond our Galaxy, in those regions of the universe that only a few decades ago appeared to us completely empty.

7 THE INVISIBLE MONSTER

At the greatest distances accessible to our telescopes, in those endless intergalactic spaces that once seemed so empty, we shall now encounter the last monster. This time, it is not an exceptional phenomenon like the anomalous red shift nor a peculiar object with well-defined characteristics, like those we have met on our journey so far. Its monstrosity is due precisely to the fact that, although we have discovered something that should exist, this something does not seem to belong to any of the categories of known objects. This mysterious intruder has been discovered in distant clusters of galaxies, but this does not mean that it can be found only within those clusters. The fact that it has been detected there is due only to the special technique responsible for its discovery. What has been discovered is that clusters of galaxies must contain masses corresponding to objects which may be different from all others known to us. As we said, however, there is no reason to believe that these objects exist only in distant clusters. On the contrary, they must also be present in the local cluster to which our Galaxy belongs, and we could conceivably expect to find them close to the solar system and perhaps to our own planet.

We should start from the facts.

When we leave the confines of our Galaxy and abandon the world of stars, our first impression is one of empty space. Facing away from the glowing Galaxy we have just left behind, we are overcome by an oppressive feeling of emptiness caused by the dark void in which our eyes strain to find something to grasp. At last we perceive a hint of light, a small luminous puff that stands out from the deep darkness. With the aid of a powerful telescope we discover that it is another galaxy, as large as ours and equally rich in stars, nebulae, and dark matter. This vast system appears like a speck of light barely discernible to the eye because we are separated by an immense distance—more than 2 million light-years. From that distance, on the other hand, our Galaxy would appear just as small and faint. As we probe deeper into the darkness with our telescope, we discover other galaxies, as far from the first two as these are from each other. We also find a number of galaxies closer to us, which our eyes had not been able to see because they are smaller, not so rich in stars, and therefore much fainter. Even though we have now discovered a great many galaxies, the density of matter in extragalactic space is very much lower than in interstellar space within the Galaxy, partly because of the immense distances between galaxies and partly because the inter- galactic regions appear nearly devoid of diffuse matter such as gas and dust.

As we continue our exploration of extragalactic space, we find entire groups of galaxies: small groups of two to five components, like those we met earlier, and very large groups comprising hundreds or even thou- sands of galaxies. These groups, or clusters of galaxies, represent the largest celestial units we can comprehend, the biggest building blocks in the universe. It is still under discussion whether there are even larger units, or superclusters of galaxies. What is certain, in any case, is that some clusters are much larger than others.

There are basically two kinds of clusters: regular and irregular. In regular clusters galaxies are distributed with spherical symmetry with re- spect to the center and in greater numbers in the central region. These clusters consist almost exclusively of elliptical or S0 galaxies, most of which are very luminous, and also very rich, comprising often a few thousand galaxies. One of the best-known clusters of this type is found in the constellation Coma Berenices (figure 7.1). It is 300 million light-years away, and 1,656 galaxies have been counted in it. Irregular clusters con- sist instead of galaxies of all types (elliptical, spiral, and irregular), dis- tributed here and there at random, without any particular concentra- tion. They are not as rich and often exhibit minor subgroups of a few galaxies each. Such is the cluster in the constellation Hercules (figure 7.2) which is also at a distance of about 300 million light-years.[1]

Zwicky and his collaborators searched for and photographed all the clusters of galaxies located in the portion of the sky observable from Mount Palomar Observatory. For this survey they used the 122-cm Schmidt telescope, which enabled them to detect clusters as far away as 2 billion light-years. In all, they found 9,700 of them. Considering that each cluster consists on the average of about 200 galaxies, we find that

---

1. In addition to this classification by G. O. Abell (1965, 1970), there are different and more complex ones proposed by various astronomers, notably, those of F. Zwicky, E. Herzeg, and P. Wild (1961), and L. P. Bautz and W. W. Morgan (1970).

within a radius of 2 billion light-years from earth there are at least 2 million galaxies.

This number does not include most of the clusters located in the southern sky, which cannot be seen from Mount Palomar. A survey of the southern hemisphere is currently under way by means of two Schmidt telescopes, similar to the one at Mount Palomar, situated respectively in Chile (European Southern Observatory) and in Australia. Moreover, Zwicky's catalogue might not include a number of clusters of compact galaxies that are difficult to detect since galaxies of this kind can be easily mistaken for stars. This type of cluster was discovered by Zwicky himself during the survey, but it seems to be fairly rare.

All the galaxies in a cluster are at the same distance from us, except for the differences due to the size of the cluster itself, which are negligible as compared to the enormous distance. Thus, when we measure their radial velocities, we should find the same value for all. However, since each galaxy may have a motion of its own within the cluster, the measured radial velocity is not only due to the cosmological effect (which increases with distance); the motion of an individual galaxy will be added to or subtracted from the velocity due to the general expansion of the universe. By measuring to what extent the radial velocities measured for the individual galaxies scatter about the mean value, we can determine how the galaxies move within the cluster, at least with respect to the component of the velocity in the line of sight.

The study of the residual radial velocities immediately appeared of the greatest interest. The motion of bodies in a system is governed by the law of universal gravitation and is determined by the mass of the system, in this case the mass of the cluster of galaxies. Thus, knowing the residual radial velocities of the component galaxies, it is possible to derive the mass of the entire cluster.

About twenty years ago calculations were performed for some of the clusters that were easier to measure, such as Coma, and their masses were obtained. The total mass of a cluster could also be determined by measuring the masses of the component galaxies and adding them up. This procedure is possible because the mass of each galaxy can be de-

Figure 7.1   The cluster of galaxies in Coma Berenices, 300,000,000 light-years away, in a photograph obtained by F. Bertola with the 122-cm Schmidt telescope. (The Hale Observatories.)

Figure 7.2   The cluster of galaxies in Hercules, about 300,000,000 light-years from us, photographed with the 5-m telescope at Mount Palomar. (The Hale Observatories.)

rived from measurements of its rotational speed obtained through spectroscopic observations.

The next step was a comparison between the values obtained with the two different methods. The result was most disconcerting because the two values of the total mass did not coincide: each cluster turned out to be heavier than all of its galaxies put together. Moreover, the difference was not a matter of a small percentage, but of several times. In the Coma cluster, for instance, the total mass was seven times larger than that obtained by summing the masses of the observed galaxies. In other words, it is as if, in addition to the number of galaxies we see in figure 7.1, the cluster contained seven times as many galaxies that we do not see. Although in other cases the difference is not so dramatic, it is now certain that clusters of galaxies are as a rule from two to ten times more massive than they should be.

At first it was thought that the real mass of the cluster was the value obtained by summing the masses of the individual galaxies. Given the observed distribution of the velocities, the consequence of this interpretation was that the galaxies would draw farther and farther apart, and the cluster would eventually break up. On the other hand, in order to keep the cluster stable and reach an agreement with the results obtained from the measured dispersion velocities, one had to assume that there was an undetected mass capable of binding the cluster together. At first, as we said, many astronomers opted for the first solution, which was consistent with the concepts of instability and expansion developed in 1954 by the Soviet astronomer V. A. Ambartsumian and already successfully applied to stellar associations. In this hypothesis, however, a cluster of galaxies should disperse in a much shorter time than the age of the universe. If the Coma cluster had a total mass equal only to the sum of its component galaxies, it should break apart in a period of about a billion years. Since the cluster exists today, its age would have to be less than a billion years. This means that it would have formed from galaxies and stars born long after the beginning of the universe. In principle this is not impossible, but it goes against observational evidence. Spectroscopic analysis of the component galaxies shows that they consist of old stars, whose ages are of the order of 10 billion years.

Similar results have been obtained for the other clusters. Hence we must conclude that clusters do not break apart, at least not so rapidly, and that the observed distribution of the velocities can be explained only by assuming that there is a hidden mass distributed in a group of objects that so far has eluded us.

We are now faced with a new problem: Where is the mass, and what does it correspond to? An answer to the first question seems to be at hand. There are only two places where the missing mass can be found: the galaxies themselves and the intergalactic spaces within the clusters. The first possibility can almost certainly be excluded. We find that the ratio between the value of the mass computed as a whole and the value obtained by adding up the individual masses increases regularly from the small groups of a few galaxies to the larger and richer clusters. A study of this phenomenon on the basis of some well-studied examples led L. M. Ozernoi to conclude that the hidden mass must be distributed for the most part in intergalactic space. This conclusion is rather attractive also from the cosmological point of view. Most likely, the process of galaxy formation, like most other natural processes, was not 100 percent efficient, and some portion of the primordial material remained unused. Assuming that the clusters are the regions in which the galaxies formed, it is reasonable to think that the leftover matter is still present between the galaxies.

In this case it should be possible to detect it in some manner. Obviously one would have to use the tools and techniques best suited to the job of detecting this material in whatever form it may be there.

First of all let us assume that the hidden mass is distributed in a large amount of intergalactic gas. This gas can be detected with different techniques depending its temperature. If it is cold, as is the interstellar gas in our Galaxy, it should emit at radio wavelengths; if it is warm, it should radiate in the visible; lastly, if it is very hot, it should emit X rays. Unfortunately, neither radio nor optical telescopes have ever revealed emissions from the Coma cluster, which has been particularly well studied. There was some hope when an X-ray emission was detected, but a computation of the mass of the gas producing the emission resulted in a value too low to solve our problem.

In late 1973 on the basis of the most reliable observations of the Coma cluster in radio, ultraviolet, and X ray, A. Davidsen, S. Bowyer, and W. Welch concluded that the extra mass that keeps this cluster bound could not be ionized gas. Neutral hydrogen was in turn excluded a year later by D. S. De Young and M. S. Roberts on the basis of radio observations in the 21-cm wavelength. Moreover, T. W. Noonan had already shown in 1972 that the missing mass could not be molecular hydrogen in intergalactic space.

Once we eliminate gas, we could think of dust. If the hidden mass was distributed in huge amounts of dust of the kind that exists between the stars of our Galaxy, the central regions of the clusters, where the observed thickness is greatest, should appear different from the limb. More accurately, the galaxies near the center should be fainter and more reddened, since the light they emit would have to traverse a greater depth and would therefore be absorbed by a larger amount of dust. These effects, which should be easily observable, have never been found. Next, we could think of larger dust particles than those observed in the Galaxy, which would cause a lower absorption per unit mass; but then we could not explain why there should be large-diameter particles without a much greater number of smaller ones.

If we now consider even larger bodies, we come to stars. Stars could be concentrated in small galaxies or in faint haloes like those discovered by H. C. Arp and F. Bertola around certain giant elliptical galaxies. Alternatively, they could be dispersed in large numbers all through the intergalactic regions of the clusters. Although the search for these objects has been successful, the mass they contribute to the system has turned out to be extremely small.

Since the hypothesis that the missing mass may be present in clusters in a form known to us has not been confirmed by observational evidence, we are forced to turn our search in other directions. We can speculate, for example, that the hidden mass may be found in black holes. As we have seen, their foremost characteristic is precisely to be invisible no matter how we look at them, and therefore they could only be detected by their mass. We now discover that in some regions of space there are mysterious masses corresponding to objects that no tool or technique currently available has been able to detect. Thus it would appear that we have found the ideal solution, and at the same time an additional indirect proof of the existence of black holes.

Although this possibility cannot be excluded, it is open to some criticism. To begin with, it seems at least imprudent to explain a phenomenon by means of objects whose existence has not yet been proved. Furthermore, in order to satisfy the required amount of mass, there would have to be an enormous number of black holes. Let us assume that their mass is on the average three times that of the sun and that the excess mass in a cluster is on the average a factor 3. Let us also assume that the solar mass corresponds to the mean value of stellar masses, so that the value that gives us the mass of a galaxy or a cluster in solar masses will also give us roughly the number of stars in the galaxy or cluster. In this hypothesis, the number of black holes required to explain the hidden mass would have to be equal to the number of stars. Moreover, if there were black holes with masses smaller than $3\,\mathcal{M}_\odot$, which could easily happen, the number would have to be even higher. Despite these difficulties, black holes still represent a possible solution.

Another possibility is that there might be dark bodies of small mass that have not been able to become stars. As we know, stars form from clouds of gas and dust that while contracting transform into heat the kinetic energy of the infalling material. Through this continuous heat production, the temperature of the body increases, and, when the inner regions reach a few million degrees, the energy-producing thermonuclear reactions are ignited. At this point the star stops contracting and remains at the same temperature, while continuously radiating energy into the surrounding space through combustion of the lighter elements. Lithium, beryllium, and boron, for which thermonuclear reactions start between 2 and 7 million degrees, burn first; then it is the turn of hydrogen, which is the most abundant element and supports the life of the star all through the long time period it is on the main sequence.[2]

In 1963 S. S. Kumar demonstrated theoretically that, when a star forms from a contracting cloud, there is a limiting value of the mass

2. See *Beyond the Moon*, pp. 212–213, 323.

below which the star cannot reach the main sequence because the temperature and density at the core are too low for hydrogen burning. The critical values are 0.07 $M_\odot$ for population I stars and 0.09 $M_\odot$ for population II stars. If the star being born from a condensing cloud has a smaller mass, it will not be able to start the hydrogen reaction as it passes through the phase of maximum core temperature and will keep on contracting until all its matter becomes degenerate, like that of a white dwarf.[3] In this instance, the temperature is lower and keeps on decreasing; so the object will not emit any light, like Sirius's companion, and will become increasingly colder and darker. Therefore a star that forms from a cloud less massive than 0.07 $M_\odot$ does not follow the normal evolutionary path through the main sequence but transforms itself into a dark object consisting of completely degenerate matter. Such objects are called "black dwarfs."

According to Kumar, it takes a forming star a much shorter time to become a black dwarf than to reach the main sequence in the course of normal evolution. A star of mass equal to the critical value of 0.07 $M_\odot$ would complete this transformation in 500 million years. A star of this mass would have a radius about one-tenth that of the sun and a maximum core temperature around 2 million degrees. This temperature might be sufficient to start the thermonuclear reactions that produce energy by burning deuterium, lithium, beryllium, and boron. During this process the contraction would stop, and the time the star needs to become a black dwarf would increase by the entire period spent in burning the lighter elements. However, Kumar found that even in this case the time would at most double. Thus it would never take longer than a billion years for a star to become a black dwarf. This means that from the beginning of the universe (about 18 billion years ago) to the present a great number of small mass stars may have become black dwarfs. Therefore the number of such objects in the universe could be very large, actually far greater than the number of the normal stars we can see.

In 1975 W. Mc D. Napier and B. N. G. Guthrie further elaborated on

3. See *Beyond the Moon*, p. 110.

these results and suggested that the undetected mass in clusters of galaxies might correspond to black dwarfs. According to these authors, a vast number of stars with masses ranging from 1/10 to 1/100 that of the sun would have formed during the development of proto-galaxies. These stars, which have never been as luminous as the sun, would still be scattered in intergalactic spaces in huge numbers. The Coma cluster, whose mass is $33 \times 10^{14} \, \mathcal{M}_\odot$ when measured as a whole, and only 1/7 of this value when computed by summing the individual galaxies, would contain over 28 billions of billions of billions black dwarfs.

This solution is also possible. But at this point we begin to realize that the only two interpretations which have not been disproved—black holes and black dwarfs—involve objects for which observational evidence has never been found in either clusters or our own Galaxy. In effect, no black dwarfs have ever been detected even in the vicinity of the solar system, where observation is easiest unless we accept Kumar's suggestion that the black bodies discovered astrometrically near a group of seven nearby stars are black dwarfs. The only advantage these objects have over all the other types of matter we have vainly searched for in clusters is that they are known from the start to be unobservable. This, however, is a two-sided argument: while it affords greater freedom for theoretical speculation, it also denies the possibility of any observational test, which after all is the only way of reaching a definite conclusion.

Thus the problem of the missing mass is still wide open. Perhaps the solution will not be as clear-cut as we have implied in our discussion. It may be, for example, that the hidden mass does not correspond to any one type of object but is distributed in different measure among gas, faint stars, black dwarfs, black holes, and so on. It is also possible that the solution will be found in phenomena we have not even considered, such as mass loss from clusters of galaxies through gravitational waves, which would invalidate the method of computing the total mass from the residual radial velocities.

Until such time that sound evidence is found for any one of these possibilities, the problem of the missing mass will remain unresolved and will be a serious limitation to our understanding of the universe.

During thousands of years of speculation and a few centuries of great scientific advances, we have come to know the planet on which we dwell and the planetary system it belongs to. We have learned that the sun governs the motion of the planets and is our source of heat and light. We have also learned that our sun is a star like billions of others, which in turn may be endowed with planetary systems. All these stars form a vast system, the Galaxy, in whose depths we have discovered many strange objects: huge, evanescent, gaseous nebulae and stars of degenerate matter like the white dwarfs; dark and insubstantial interstellar dust and super-dense neutron stars; associations where stars are born and pulsars, which are the last remnants of dead stars; isolated stars that spinning furiously surround themselves with gas rings and multiple stars which in rotating rapidly about one another exchange mass and are continuously transformed. Exploring space outside the Galaxy, we have learned that all we know about one galaxy, ours, is also true for the millions of others that populate the universe. In addition, we have found new and extraordinary objects: Lacertids, Markarian galaxies, and quasars. The number and variety of objects currently known in the cosmos are very hard to grasp even for the scientists who have discovered and studied them.

Nevertheless, if the mysterious masses detected in clusters of galaxies really exist, it means that even in the part of the universe accessible to us we can explain—and roughly at that—only a third of what is actually there. The rest is mystery.

APPENDIX: RED SHIFT AND DISTANCE

In chapter 6 we sometimes expressed the distance of galaxies r in Mega-parsec or in millions of light-years and alternately with the red shift z shown by their spectra or with the recession velocities corresponding to such red shifts. For the farthest objects, however, we used almost exclusively the value z, as is customary for astronomers. The reasons for this are neither capricious nor faddish.

The use of red shifts to measure distances became prevalent in the 1930s after the discovery of Hubble's law which relates the two quantities, z and r. Note that $z = \Delta\lambda/\lambda$ is the value of the measured red shift. Assuming that the red shift is due to the Doppler effect, then

$$z = \frac{v}{c}, \tag{1}$$

where v is the recession velocity of the object and c, as usual, the speed of light. From the recession velocity v the distance of the object in Megaparsec can be computed through the formula

$$r = \frac{v}{H}, \tag{2}$$

where H is Hubble's constant (its most current value is $H = 55$ km/s/Mpc, as given by Sandage in 1975). To obtain the distance in light-years recall that 1 Mpc = 3,260,000 light-years.

These formulae hold only until $z \leq 0.1$, or until $v \leq 30,000$ km/s. For higher values the relativistic formula,

$$z = \sqrt{\frac{c + v}{c - v}} - 1, \tag{3}$$

must be used to derive v from z, yielding

$$v = \frac{2cz + cz^2}{2 + 2z + z^2}.$$

This formula gives the correct value of v, but to find the distance r, we can no longer use equation (2). To obtain r we must apply the formula,

$$z = \frac{Hr}{c} + \frac{1}{2}(1 + q_0)\left(\frac{Hr}{c}\right)^2 + \text{higher-order terms;} \tag{4}$$

note that equation (1), a first approximation applicable for $z \leqslant 0.1$, takes into account only the first-term expression in (4) [from (2) it follows that $v = Hr$]. Equation (4) takes into account also the form factor of the universe (introduced in the second term through q_0, the deceleration parameter) which becomes important when the measurement is performed at very large distances, at $z > 0.1$.

Unfortunately, thus far the measurement of q_0 has given scientists uncertain and contradictory results. This means that they have not been able to determine from observations just what type of universe we live in—whether it is closed, open, or flat. Hence the distance of objects farther away than 3 billion light-years cannot be exactly computed unless a priori a certain model of the universe is assumed. Therefore distances derived from equation (4) for galaxies and quasars farther away than 3 billion light-years will be affected by an error that increases the larger the distance and the larger the discrepancy between the assumed q_0 and the unknown true one. Another problem is that for $z \geqslant 1$ equation (4) becomes invalid. This is the reason why astronomers simply report the observed red shift value z for the most distant objects. It is only from a comparison of this z with the z values of the closer objects that some indication of distance can be obtained for those galaxies farthest from us. For two objects with different z values the closer will have the smaller z.

BIBLIOGRAPHY

CHAPTER 1: THE FASCINATING MONSTERS

Sekanina, Z. *Disintegration phenomena in comet West.* In Sky and Telescope, **51**, 386 (1976).

Jacchia, L. *The brightness of comets.* In Sky and Telescope, **47**, 216 (1974).

Whipple, F. L. *La natura delle comete.* In Le Scienze (May 1974).

Tempesti, P. *Parliamo di comete.* Teramo (1973).

Chebotarev, G. A., Kazimirchak-Polonskaya, E. I., Marsden, B. G. *The motion, evolution of orbits and origin of comets.* Dordrecht (1972).

Marsden, B. G. *Catalogue of cometary orbits.* Cambridge (1972).

Kazimirtchak-Polonskaya, H. I. *Rôle des planètes extérieurs dans l'évolution des orbites des comètes.* In L'Astronomie, **82**, 217, 323, and 432 (1968).

Marsden, B. G. *The sungrazing comet group.* In Astron J., **72**, 1170 (1967).

Krinov, E. L. *Giant meteorites.* Oxford (1966).

Vsekhsvyatskii, S. K. *Physical characteristics of comets.* Jerusalem (1964).

Richter, N. B. *The nature of comets.* London (1963).

Wurm, K. *The physics of comets.* In *The solar system*, IV. Chicago (1963).

CHAPTER 2: PHANTOMS OF THE SOLAR SYSTEM

Combes, M. A. *Les variations orbitales de la comete de Halley.* In L'Astronomie, **87**, 103 (1973).

Fonteurose, R. *In search of Vulcan.* In J. for Hist. of Astron., **4**, 145 (1973).

Kiang, T., Wayman, P. A. *The orbit of Halley's comet.* In Nature, **241**, 520 (1973).

Rawlins, D., Hammerton, M. *Mass and position limits for a hypothetical tenth planet of the solar system.* In Mont. Not. R. Astr. Soc., **162**, 261 (1973).

Brady, J. L. *The effect of a trans-plutonian planet on Halley's comet.* In Pub. Astr. Soc. Pacific, **84**, 314 (1972).

Seidelmann, P. K., et al. *Note on Brady's hypothetical trans-plutonian planet.* In Pub. Astr. Soc. Pacific, **84**, 858 (1972).

Kiang, T. *The past orbit of Halley's comet.* In Mem. R. Astr. Soc., **76**, 27 (1971).

O'Keefe, J. A. *Tektites from Tycho?* In Sky and Telescope, **38**, 389 (1969).

Arend, S. *Saturne n'est pas la seule planète à avoir un anneau, la Terre a aussi le sien.* In Ciel et Terre, **83**, 132 (1967).

O'Keefe, J. A. *Tektites.* Chicago (1963).

O'Keefe, J. A. *Tektites and the Cyrillid shower.* In Sky and Telescope, **21,** 4 (1961).

Eggen, O. J. *Vulcan.* In Astron. Soc. Pacific, Leaflet n. 287 (1953).

CHAPTER 3: THE KILLER MONSTERS

Miller, J. S. *La struttura delle nebulose a emissione.* In Le Scienze (January 1975).

Miller, E. W., Muzzio, J. C. *Mosaics of a southern nebula.* In Sky and Telescope, **49,** 94 (1975).

Cosmovici, C. *Supernovae and supernovae remnants.* Dordrecht (1974).

Lindblad, P. O. *Gould's Belt.* In Highlights of Astronomy, **3,** 381 (1974).

Weaver, H. *Space distribution and motion of the local HI gas.* In Highlights of Astronomy, **3,** 423 (1974).

van den Bergh, S., et al. *An optical atlas of galactic supernova remnants.* In Ap. J. Suppl., **26,** 227 (1973).

Hughes, V. A., Routledge, D. *An expanding ring of interstellar gas with center close to the Sun.* In Astron. J., **77,** 210 (1972).

Maran, S. P. *La nebulosa di Gum,* in Le Scienze (March 1972).

Bok, B. J. *The Gum Nebula.* In Sky and Telescope, **42,** 64 (1971).

Gorenstein, P., Tucker, W. *I resti delle supernovae.* In Le Scienze (October 1971).

Maran, S. P., et al. *The Gum Nebula etc.* In Physics Today (September 1971).

Gurzadyan, G. A. *Planetary nebulae.* Dordrecht (1970).

Osterbrock, D. E., O'Dell, C. R. *Planetary nebulae.* Dordrecht (1968).

Shklovski, I. S. *Supernovae.* London (1968).

Struve, O. *Runaway Stars.* In Sky and Telescope, **21,** 140 (1961).

CHAPTER 4: ETA CARINAE

ETA CARINAE
Deharveng, L., Maucherat, M. *Optical study of the Carina Nebula.* In Astron. Astrophys., **41,** 27 (1975).

Hutchmeier, W. K., Day, G. A. *The Carina Nebula at 3.4 and 6 cm.* In Astron. Astrophys., **41,** 153 (1975).

Walborn, N. R. *Forbidden S II structures in the Carina Nebula.* In Ap. J. Letters, **202,** L129 (1975).

Feinstein, A., Marraco, H. G. *On a possible three year cycle of Eta Carinae.* In Astron. Astrophys., **30,** 271 (1974).

Gehrz, R. D., et al. *The infrared spectrum and angular size of Eta Carinae*. In Astrophys. J., letters, **13,** 89 (1973).

Neugebauer, G., Becklin, E. E. *Le sorgenti infrarosse più brillanti*. In Le Scienze (July 1973).

Gehrz, R. D., Ney, E. P. *The core of Eta Carinae*. In Sky and Telescope, **44,** 4 (1972).

Gardner, F. F., et al. *The Carina Nebula at 6 cm*. In Astron. Astrophys. **7,** 349 (1970).

Westphal, J. A., Neugebauer, G. *Infrared observations of η Car to 20 μm*. In Astrophys. J., letters, **156,** L45 (1969).

Gerola, H., Viotti, R. *Sulla distanza di Eta Carinae*. In Mem SAIt, **38,** 451 (1967).

Gratton, L. *The problem of Eta Carinae*. In *Rendiconti Scuola Internazionale di Fisica "Enrico Fermi," Varenna XXVIII Corso*. New York (1963).

Gratton, L. *L'evoluzione del mondo inorganico*. In La Ricerca Scientifica, **1,** 50 (1961).

Gaviola, E. *Eta Carinae—The spectrum*. In Astrophys. J., **118,** 234 (1953).

Thackeray, A. D. *Identifications in the spectra of η Car and RR Tel*. In Mon. Not. R. Astr. Soc. **113,** 211 (1953).

Vancouleurs, G. de. *The wonder star: Eta Carinae*. In Astron. Soc. Pacific, Leaflet n. 281 (1952).

Bertaud, Ch. *Une étrange nova: Eta Caréne*. In L'Astronomie, **65,** 343 (1951).

SIMILAR STARS
Bianchini, A., Rosino, L. *The spectrum of the bright var A-1 in M 31*. In Astron. Astrophys., **42,** 289 (1975).

Humphreys, R. M. *The spectra of AE And and the Hubble-Sandage variables in M 31 and M 33*. In Astrophys. J., **200,** 426 (1975).

Rosino, L., Bianchini, A. *Observations of Hubble-Sandage variables in M 31 and M 33*. In Astron. Astrophys., **22,** 453 (1973).

Tammann, G. A., Sandage, A. *The stellar content and distance of the galaxie NGC 2403 in the M 81 group*. In Astrophys. J., **151,** 825 (1968).

Hoffmeister, C. *Die veränderliche Sterne*. Lipsia (1968).

Bertola, F. *The supernovae in NGC 1073 and NGC 1058*. In *Novae, Novoïdes et Supernovae*. Paris (1965).

Zwicky, F. *NGC 1058 and its supernova 1961*. In Astrophys. J., **139,** 514 (1964).

Bertaud, Ch. *R R Telescopii, curieuse nova du ciel austral.* In L'Astronomie, **74,** 121 (1960).

Hubble, E., Sandage, A. *The brightest variable stars in extragalactic nebulae; M 31 and M 33.* In Astrophys. J., **118,** 353 (1953).

Bertaud, Ch. *Une nova a très lente évolution.* In L'Astronomie, **65,** 7 (1951).

Bertola, F., Arp, H. *NGC 1058 and its peculiar supernova.* In Pub. Astr. Soc. Pacific, **82,** 894 (1970).

CHAPTER 5: BLACK HOLES

Avni, Y., Bahcall, J. N. *Ellipsoidal light variations and masses of X-ray binaries.* In Astrophys. J., **197,** 675 (1975).

Block, D. L. *Black holes and their astrophysical implications.* In Sky and Telescope, **50,** 20, and 87 (1975).

Chapline, G. F. *Cosmological effects of primordial black holes.* In Nature, **253,** 251 (1975).

Gursky, H., van den Heuvel, E. P. J. *Sorgenti di raggi X in sistemi binari.* In Le Scienze (July 1975).

Thorne, K. S. *La ricerca dei buchi neri.* In Le Scienze (April 1975).

Chandrasekhar, S. *Development of general relativity.* In Nature, **252,** 15 (1974).

Griblin, J. *Retarded cores, black holes and galaxy formation.* In Nature, **252,** 445 (1974).

Kaufmann, W. J. *Pathways through the universe—Black holes, worm holes, and white holes.* In Mercury, **3,** 26 (1974).

Rees, M. J. *Black holes.* In Observatory, **94,** 168 (1974).

Misner, C. W., Thorne, K. S., Wheeler, J. A. *Gravitation.* San Francisco (1973).

Ruffini, R. *Relatività generale e fisica della Galassia.* In *Scienza & Tecnica 73.* Milan (1973).

Penrose, R. *I buchi neri.* In Le Scienze (August 1972).

Schopper, H. W., Delvaille, J. P. *Il cielo ai raggi X.* In Le Scienze (October 1972).

Ruffini, R., Wheeler, J. A. *Introducing the black hole.* In Physics Today, **24,** 30 (1971).

Lynden-Bell, D. *Galactic nuclei as collapsed old quasars.* In Nature, **223,** 690 (1969).

Novikov, I. D. *Delayed explosion of a part of the Fridman universe, and quasars.* In Soviet Astronomy, **8,** 857 (1965).

Ne'eman, Y. *Expansion as an energy source in quasi stellar radio sources.* In Astrophys. J., **141,** 1303 (1965).

CHAPTER 6: THE KEY MONSTERS

LACERTIDS
Baldwin, J. A., et al. *The nature of BL Lacertae.* In Astrophys. J., letters, **195,** L55 (1975).

Carswell, R. F. *Red shifts of BL Lac objects.* In Nature, **253,** 589 (1975).

Kinman, T. D. *The surface brightness of the nebulosity in BL Lacertae.* In Astrophys. J., letters, **197,** L49 (1975).

Liller, M., Liller, W. *Photometric histories of QSOs: two QSOs with large light amplitude.* In Astrophys. J., letters, **199,** L133 (1975).

Thuan, T. X., Oke, J. B., Gunn, J. E. *Further observations of BL Lacertae.* In Astrophys. J., **201,** 45 (1975).

Ulrich, M. H., et al. *Nonthermal continuum radiation in three elliptical galaxies.* In Astrophys. J., **198,** 261 (1975).

Oke, J. B., Gunn, J. E. *The distance of BL Lacertae.* In Astrophys. J., letters, **189,** L5 (1974).

Shaffer, D. B., Shields, G. A. *A search for additional radio sources in the Kukarkin variable star catalog.* In Astrophys. J., letters, **192,** L83 (1974).

Shapiro, S. L., Elliot, J. L. *Are BL Lac-type objects nearly black holes?* In Nature, **250,** 111 (1974).

Stull, M. A. *Two puzzling objects: OJ 287 and BL Lacertae.* In Sky and Telescope, **45,** 224 (1973).

Hackney, R. L., et al. *Optical and radio flares in BL Lacertae (VRO 42.22.01).* In Astrophys. J., letters, **12,** 147 (1972).

Strittmatter, P. A., et al. *Compact extragalactic nonthermal sources.* In Astrophys. J., letters, **175,** L7 (1972).

Andrew, B. H., et al. *OJ 287: An exceptionally active variable source.* In Astrophys., letters, **9,** 151 (1971).

Kinman, T. D., Conklin, E. K. *Observations of OJ 287 at optical and millimeter wavelengths.* In Astrophys., letters, **9,** 147 (1971).

MacLeod, J. M., et al. *Comprehensive observations of the rapidly varying radio source VRO 42.22.01 (BL Lac).* In Astrophys., letters, **9,** 19 (1971).

Blake, G. M. *Observations of extragalactic radio sources having unusual spectra.* In Astrophys., letters, **6,** 201 (1970).

MARKARIAN GALAXIES
Sramek, R. A., Tovmassian, H. M. *A radio survey of Markarian galaxies at 6 centimetres.* In Astrophys. J., **196,** 339 (1975).

Khachikian, E. Ye., Weedman, D. W. *An Atlas of Seyfert galaxies.* In Astrophys. J., **192,** 581 (1974).

Markarian, B. E. *The nature of galaxies with ultraviolet continuum—Principal spectral and color characteristics.* In Astrophysics, **8,** 100 (1974).

Huchra, J., Sargent, L. W. *The space density of the Markarian galaxies including a region of the south galactic hemisphere.* In Astrophys. J., **186,** 433 (1973).

Weedman, D. W. *A photometric study of Markarian galaxies.* In Astrophys. J., **183,** 29 (1973).

Weedman, D. W. *Emission-line intensities and UBV magnitudes for 23 Markarian galaxies.* In Astrophys. J., **171,** 5 (1972).

II catalogo delle galassie di Markarian è stato pubblicato in cinque elenchi contenuti. In Astrofizika, **3,** 55 (1967); **5,** 443 (1969); **5,** 581 (1969); **7,** 511 (1971); **8,** 155 (1972).

N GALAXIES
Kukarkin, B. V., et al. *Catalogo generale delle stelle variabili.* 2d. suppl., 3rd ed., 347 (1974).

Sandage, A. *The composite nature of N galaxies, their Hubble diagram, and the validity of measured red shifts as distance indicators.* In Astrophys. J., **180,** 687 (1973).

Burbidge, G. R. *The nuclei of galaxies.* Erice (1971).

Sargent, W. L. *Compact galaxies.* Erice (1971).

Sargent, W. L. *The optical line and continuous spectra of radio galaxies, compact galaxies and Seyfert galaxies.* In *Les noyaux des galaxies.* Vatican City (1971).

SEYFERT GALAXIES
Adams, T. F., Weedman, D. W. *Emission-line luminosities of Seyfert galaxies.* In Astrophys. J., **199,** 19 (1975).

van den Bergh, S. *The classification of active galaxies.* In J. of R. Astr. Soc. Canada, **69,** 105 (1975).

Khachikian, E. Ye., Weedman, D. W. *An Atlas of Seyfert galaxies.* In Astrophys. J., **192,** 581 (1974).

QUASARS
Goughuenheim, L. *Quasars: le mystère s'épaissit.* In L'Astronomie, **89,** 59 (1975).

Pollock, J. T. *Variability and optical-radio properties of BL Lacertae objects*. In Astrophys. J., letters, **198**, L53 (1975).

Usher, P. T. *A comparison of some radio and optical properties of quasi-stellar sources and BL Lacertae objects*. In Astrophys. J., letters, **198**, L57 (1975).

Kristian, J. *Quasars as events in the nuclei of galaxies: the evidence from direct photographs*. In Astrophys. J., letters, **179**, L61 (1973).

ANOMALOUS RED SHIFTS
Chamaraux, P. *Le quintette de Stephan*. In L'Astronomie, **88**, 115 (1974).

Goughuenheim, L. *Les décalages spectraux "anormaux."* In L'Astronomie, **88**, 17 (1974).

The quasar controversy—an interview with Caltech astronomer Halton Arp. In Mercury, **3**, 8 (1974).

Tribunal des red shift anormaux. Meudon (1973).

Field, G. C., Arp, H., Bahcall, J. N. *The red shift controversy*. Reading (1973).

Arp, H. *Atlas of peculiar galaxies*. In Astrophys. J., suppl., **14** (1966).

CHAPTER 7: THE INVISIBLE MONSTER
Margon, B. *The missing mass*. In Mercury, **4**, 2 (1975).

Napier, W., Guthrie, B. *Black dwarf stars as missing mass in clusters of galaxies*. In Mont. Not. R. Astr. Soc., **170**, 7 (1975).

Ozernoi, L. M. *Where is the "hidden" mass?* In Sov. Astr., **18**, 654 (1975).

Kumar, S. S. *Planetary systems*. In *The emerging universe*. Charlottesville (1972).

INDEX